ENVIRONMENTAL HEROES

Success

Stories

of People

at Work for

the Earth

KEVIN GRAHAM
GARY CHANDLER

PRUETT PUBLISHING COMPANY
BOULDER, COLORADO

Printed in the United States
10 9 8 7 6 5 4 3 2 1

Library of Congress Cataloging-in-Publication data

Graham, Kevin, 1959–
        Environmental heroes : success stories of people at work for
the earth / Kevin Graham and Gary Chandler.
            p.   cm.
        Includes index.
        ISBN 0-87108-866-5 (pb)
            1. Environmentalists—Handbooks, manuals, etc.
        2. Environmentalism—Handbooks, manuals, etc. I. Chandler,
Gary.
        II. Title.
        GE55.G75   1996
        363.7'0092'2—dc20   96-32839
                                                                    CIP

COVER AND INTERIOR BOOK DESIGN:
Rebecca Finkel, F + P Graphic Design

To our parents

CONTENTS

CONTENTS

Acknowledgments

The authors thank Scott Graham for his contribution to this book. He kept us on course with his experience and keen editor's eye. We also want to recognize the countless environmental efforts not mentioned in this book. Most of them rarely get the recognition they deserve.

How We Got From There to Here: The Evolution of Environmental Awareness

A river burns in Ohio. Bridges—not usually considered fire hazards—are threatened by the inferno, which is fueled by the chemicals and wastes that have been dumped into the Cuyahoga River. It is the late 1960s, and America has come face to face with its mounting environmental problems. Life-forms in lakes are dying or bringing malformed young into the world due to an abundance of toxic pollution. The air in many U.S. cities is so dirty that residents finally are realizing something needs to be done.

Fortunately, many people and organizations at the time dedicated themselves to protecting the environment. Even before this culmination of ecological crises, groups of people had been working to ensure a safe environment in the future. For example, the Sierra Club, then already more than seventy-five years old, was battling to preserve wilderness and halt pollution. In 1968, the club helped establish North Cascades and Redwood National Parks, as well as the federal Wild and Scenic Rivers System.

Thousands more efforts have been undertaken since—all successes in their own right. This book showcases a number of these environmental success stories. They cover a wide spectrum of endeavors—from reusing old computer equipment to building rock pyramids as a way of producing fresh water, from saving Africa's endangered mountain gorillas to composting waste material created by zoo animals. These success stories have been divided into six sections, each covering topics under a central theme, such as wildlife, recycling, or alternative energy. In describing these efforts, this book offers readers ways to understand, appreciate, and help the environmental movement. It provides:

- Practical knowledge about measures and projects now under way to protect the environment. Phone numbers and addresses listed at the end of each chapter give readers the opportunity to get involved in any number of ways.
- A basic understanding of the complicated nature, wide variety, and severity of the many environmental issues human society faces today. Many

factors lead to modern environmental problems, including such issues as poverty, politics, and social injustice.

• A good sense of the refreshing diversity to be found in the people profiled and interviewed for this book. Some are ordinary folks who had an idea, ran with it, and created a successful environmental endeavor. Others have put their hard-earned educational and professional backgrounds to use for the environment.

But before we discuss these encouraging achievements in greater detail, one question deserves an answer: How did the environment come to be such a crucial issue in the first place?

Technically, the "environment" entails everything. A stop sign is part of the environment. So, too, is a plastic Christmas tree. As we have moved toward the end of the twentieth century, however, modern society has modified the word "environment." It has come to mean strictly the natural environment—air, water, plants, animals. In a word, *nature.*

Only a few hundred years ago, protecting the natural world was not crucial to humanity's survival. With billions fewer people on the earth, there was no shortage of open land for all plants and animals to thrive, including human animals. But the Industrial Revolution of the 1800s transformed this overwhelmingly rural way of life and altered many societies. People flocked to the cities of Europe and North America to work in factories. This shift from agriculture to industry took its toll on the environment. Forests were cleared to provide wood for new buildings and paper products. Larger amounts of minerals were pulled from the earth for industrial purposes, and water use increased accordingly. Pollution, industry's infamous by-product, became one of the biggest problems— everything from wastewater to foul air to heaping piles of trash mounted as the juggernaut of the industrial age entered the twentieth century.

As time passed and industries grew, pollution became more severe, complex, and widespread. And because every person on the planet causes pollution in some form, the growing human population meant more pollution. Over the last 150 years, the earth has withstood a tremendous pounding from pollution. But it wasn't until the 1960s that people around the world began understanding the potential long-term effects of environmental pollution. When a river catches fire, people take notice.

This growing concern culminated in the first Earth Day in 1970. The event drew attention to the degraded environment and spurred scientists and governments to step up their efforts to reduce pollution and protect the natural world.

The U.S. government created the Environmental Protection Agency (EPA) later the same year. A number of federal programs attempted to control the amount of pollution caused by various industries and businesses. State and city governments followed the EPA's lead by developing their own pollution-control efforts. Gradually, a body of environmental law was enacted. This created the foundation for many of the debates over environmental issues we are witnessing today. To some people, controlling pollution and saving ecosystems means potentially reducing profits and limiting development. People on both sides of environmental issues take firm positions and square off.

Many of these controversies are discussed in this book. Are old-growth forests worth saving? The timber industry believes there are plenty of old-growth forests already set aside in parks and wilderness areas scattered throughout North America. Environmentalists, on the other hand, argue that only about 10 percent of our old-growth forests remain standing and that enough logging has already taken place in these precious forests.

Should millions of acres of undeveloped wildlands in southern Utah be protected for the enjoyment of the human population and the preservation of wildlife, or should they be developed for economic gain through increased mining, oil and gas drilling, and cattle grazing? A nonprofit effort in the state argues in favor of preserving at least a portion of this land because plenty of other acreage has already been given away to industry and plenty more would be left for future development. The land in question contains numerous irreplaceable Native American archaeological features as well as geological wonders, the group says, and the future economy of the West should be tied to these natural treasures, not to the further consumption of natural resources that destroys spectacular landscapes.

Prodevelopment forces, however, believe the land is better suited for mining and cattle ranching and should therefore be opened to such uses. More than 800,000 acres of wilderness already exist in Utah, they say, and so only a fraction of the acreage in question needs to be designated as wilderness. Opening up the rest of the land to development could help the economic plight of many rural Utah counties that now lack alternatives because more than 70 percent of the state's lands are owned, protected, or controlled in one way or another by the federal government.

Another controversy involves wind power. Is it worth pursuing this renewable energy source on a large scale? Some people would say yes, but opponents site a bird-mortality issue: Many raptors are killed each year by the spinning wind-turbine blades. For more efficient, pollution-free energy, can we spare the cost of rare birds?

On yet another front, is the ozone layer that protects the earth from some of the sun's harmful rays actually disappearing? And if it is, will it really make any difference to life on the planet?

Who's right and who's wrong when it comes to environmental questions and answers such as these? Where are the lines drawn, and who draws them?

Many prodevelopment groups and politicians say environmental concerns are not as crucial as environmentalists fear. They believe, for instance, that the earth's natural systems have plenty of healing capacity and that a few less species won't hurt anything. To many, industrial and economic growth simply are more important than open stretches of land and clean sources of air and water. In fact, clear-cut answers to many environmental questions do not exist. Science itself, unfortunately, is often of dubious value in these issues.

But even if the critics of environmental efforts prove to be right in the long run, can life on earth afford the risk of finding out that they were wrong? All of the folks in this book say no—we cannot afford the risk, the price is too high. When the ozone layer is gone, there is no replacing it. Facing this kind of finality, we believe conservation clearly is the safe and wise choice.

As all these debates continue to roil, millions of people around the world are growing increasingly concerned about the future of the environment. Many live in poorer nations and face a life of poverty with little or no control over their natural world. They watch as their environment is degraded for profit by powerful and influential segments of their own societies and of foreign industrial nations, or for lack of alternatives are forced to damage it themselves in order to survive. Norwegian Prime Minister Gro Harlem Brundtland, who started her political career as her country's environmental minister, believes that because many people in the Third World face a life of despair and hardship, they can't be expected to see environmental protection as a primary concern. Although they might want to protect the environment—such as the world's dwindling rain forests—their economic and political plight simply won't allow it.

In the Third World, often the poor have no alternatives in their daily struggle to survive other than overusing natural resources. "Alleviating poverty should be priority No. 1. Very little else will matter if more than one billion people continue to live in absolute destitution," Brundtland said in an April 1993 interview in *Technology Review* magazine. "Only by educating people and giving them a fair chance to break out of poverty can we hope to find a sustainable relationship between population and resources. Otherwise, we will be forced to continue overusing natural resources. This is what [former Indian prime minister] Indira Ghandi meant when she said, 'Poverty is the greatest polluter.'"

Faced with few employment alternatives, for instance, many of the world's poor must survive on slash-and-burn farming practices that destroy forests and deplete soil in only a few years. Eventually, the effects of their struggles will affect us all—unless we change the way we treat the world's threatened ecosystems and the people who live in and around them. Many scientists report that the destruction of the world's tropical rain forests will cause drastic climate changes around the globe. These forests absorb a huge amount of carbon dioxide, a gas that contributes to the greenhouse effect, or global warming. Scientists and others suggest this phenomenon will lead to a catastrophic environmental crisis in the future as glaciers melt, oceans rise, and growing seasons are disrupted.

But again the question remains: Whether or not critics of the environmental movement happen to be right in the long run, can we afford the risk of discovering that they are wrong? Several endeavors profiled in this book are involved with rain forest issues—new medicines are being discovered thanks to the help of native healers, and people living in or near the forests are gradually learning ways to sustain themselves and their societies without harming the ecosystem.

In the mid-1980s, Brundtland led one of the most intensive studies on the future of the global environment ever undertaken. Called the World Commission on Environment and Development, the three-year effort made the words "sustainable development" a common term. The concept entails creating economic growth without destroying or depleting the natural world. And the crux of this effort falls to the world's industrialized nations, which use a disproportionate share of the planet's resources and create a corresponding proportion of modern society's pollution—many times at the expense of developing nations. The United States, for instance, constitutes about 5 percent of the earth's population yet accounts for 25 percent of the world's energy consumption. Nonetheless, sustainable growth, if created and distributed equitably, can produce the monetary resources necessary to solve numerous environmental and social problems worldwide.

"The time has come for a marriage of economy and ecology so that governments and their people can take responsibility not just for environmental damage, but for the policies that cause the damage," the commission stated in its report, Our Common Future, also known as The Brundtland Report. "Poverty is a major cause and effect of global environmental problems."

Of course, poverty didn't cause a U.S. river to catch fire, and it didn't create the 1,800 heavily polluted Superfund sites around the United States. These toxic sites, designated for cleanup by the EPA, are the result of past industrial dumping practices and military weapons activities. Questionable government policies combined with corporate ignorance and greed in the industrial world have created major environmental problems. In too many cases the environment continues to be ignored by big business and government. And in doing so, corporations and governments exacerbate the problems Brundtland speaks out against. As she says it, it comes down to finding common ground between economic gain and environmental protection.

Despite her deep concerns about population growth, worldwide poverty, and the overall degradation of the earth's environments, Brundtland is hopeful for the future of the planet. "I am convinced that we will succeed in standing up to the dangers facing us because there are simply no alternatives," she once said. "We must manage the most important global transition since the agricultural and industrial revolutions . . . Someday, when people look back on the present generation, we want them to be able to say: Faced with the challenge, they managed to upgrade human civilization."

The environmental movement isn't just about saving whales, trees, or spotted owls—it's about saving life as we know it. Because all plants and animals on earth are interconnected through the web of life, losing any species is like cutting a strand from that web. Once enough strands are cut, the web will collapse, taking future generations down with it. Can we afford to risk these consequences?

All of the people featured in this book are working to mitigate modern society's impact on the natural world through a variety of unique and important environmental efforts. Plenty of ecological problems face our world today. These people and organizations are tackling many of them by promoting existing solutions or searching for new ones. For example, Arizona entrepreneur Sally Fox knew of the harmful effects of the dyes and chemicals used in the cloth manufacturing industry, so she began growing "colored" cotton. In Massachusetts, Will Brinton realized the environmental benefits of composting and set out to make the practice better understood and easier to undertake. And in California, Mark Hass saw thousands of old computers being hauled to a landfill and decided there had to be a way to recycle these still-usable machines for schools.

As a reader of this book, you can view these people as role models and seek to follow their examples by creating other ways to benefit the environment, or you can simply join in their efforts and help them realize their goals. If a particular effort interests you, give the organization a call and get more information—get involved and make a difference.

All the people profiled in this book are extraordinary in that they have committed their lives to helping protect the environment. They have taken their concerns and confronted the various problems facing society today through a myriad of efforts. This book celebrates those efforts. The people here have been highlighted to inspire even more positive action and to provide a sense of hope and possibility. Many of them believe humanity's future depends on a healthy and stable environment and realize that modern society is on a collision course with the environment. They are working to alter that course and to create a sustainable future.

Of course, there are thousands—indeed millions—of other people working on many fronts to clean up environmental messes or to protect natural resources. The environmentalists featured in this book join them in trying to attain Gro Harlem Brundtland's optimistic view of the future—to upgrade human civilization. And to make sure that no more rivers catch fire.

RECYCLING

The computer age has added a whole new dimension to the concept of recycling. Here, recycled computer parts make up the distinctive look of a Tecnotes clock. (Photograph courtesy of Tecnotes. Used with permission.)

CHAPTER 1

COMPUTING AWAY TRASH

The computer world is fast-paced and ever changing. New ideas and technology prompt new products—which seemingly hit the market every other day. About the time the average consumer or business learns how to use a new computer system or software package, it becomes outdated. This phenomenon spells trash, and plenty of it. However, new recycling ideas are providing alternatives for much of the outmoded equipment and peripherals used in the computer world. These ideas, incorporated into nonprofit efforts or small businesses, are allowing people to reuse older equipment instead of hauling it to landfills, where it would sit for centuries.

Recycling allows items of all kinds to find second lives. It is well known that aluminum cans are easily reprocessed into new items, saving the expense of finding and processing virgin supplies of the metal. In a similar way, a multitude of other products and materials can be recycled—and the more items that can find second or third lives, the better it is for the environment. Recycling not only keeps waste products out of our landfills but also eliminates the need to further damage the environment in search of new raw materials.

In the computer world, the conservation age has now met the information technology era, and new ideas are transforming many high-tech waste problems into social and economic opportunities. In 1991, for instance, Californian Mark Hass faced the task of clearing out a warehouse filled with old computers. He first tried to give them away. But with too little time, however, he was unsuccessful and ended up throwing them all away. That experience spawned an idea that has grown into a nonprofit corporation called the Computer Recycling Center. Hass, along with Steven Wyatt and Will Marshman, founded the organization in 1992 after discovering that many companies were happy to give away their old computers and that schools were equally happy to receive them. "The idea got an excellent reception," Hass said. "It was clear from the beginning that companies had a problem getting rid of their old computers."

With both collection and distribution systems now in place, the Computer Recycling Center takes in thousands of old computers and parts each month at its main facility in Mountain View, California—a city near the famed Silicon Valley in San Jose—as well as at two other northern California locations. Semitrailers loaded with used computer equipment arrive every week from such companies as IBM, the Bank of California, National Semiconductor, and others.

A staff of hundreds of volunteers then tears into the donated machines, readying them for use in public schools. The volunteers clean the computers and perform diagnostic tests of the various operating systems. Broken machines provide spare parts. In addition, Geoworks Corporation, a computer software firm, allows the center to place a working copy of its Geoworks Pro software on every donated computer. The software features excellent graphics and runs on older computers, Hass says, giving older machines new and useful lives.

Because the center does not pay shipping fees, schools must come to the distribution facilities to pick up computers or must arrange for shipping. Requests for computers must be for public education and made through tax-exempt, nonreligious organizations. Now in its fourth year of operation, the center has placed tens of thousands of computers, printers, monitors, sealed software packages, and other peripherals in California schools and nonprofit organizations. "We're now in the process of replicating our operation—helping get other centers started," Wyatt says. "We have state officials, city managers, community leaders, and entrepreneurs from across the country visiting us and requesting our help in setting up local and regional centers elsewhere."

Without suitable alternatives, more than 6 billion pounds of computer hardware will end up in landfills over the next ten years. There are an estimated 60 million personal computers currently operating in the United States, each with an average useful life of five years. "In every major city, each of those big buildings has a computer sitting on every desk—and all companies periodically upgrade them and have to do something with the old computers," Mark Hass explains. "We're setting up a conduit to create a flow of computers to schools. It's good for the community, businesses, and the environment."

In addition to its computer recycling efforts, the center is a state-licensed postsecondary vocational training school. "We've developed a variety of courses involving hardware, software, applications, and networks, as well as an internship program in these areas," Hass says. "We target displaced and disabled workers needing retraining, along with junior high and high schools that want to develop work programs for their students in computer repair and network administration."

Another of the center's ideas tackles the crime issue. At a 1995 guns-for-computers swap, hundreds of sawed-off shotguns, handguns, and other firearms were turned in for used computers on a one-for-one basis. Because of the idea's initial success, more swaps are planned. Yet another idea involves opening up "street schools" in empty storefronts by lining up used computers for anyone to use and learn on. "With the help of volunteers at these street schools," Hass says, "anyone will be able to walk in and learn how to use computers to either boost their job skills or simply their self-esteem."

In another recycling effort, used computers are entering the art world. When Luca Bonetti came across a bunch of circuit boards that had been removed from old computers, he didn't see trash, he saw art. As a result, he started a company that recycles the old boards into a number of unique products.

Circuit boards are a vital component of nearly every modern electronic product. They are durable plastic sheets, stamped with intricate copper mazes and multiple series of holes where transistors are eventually placed, which fit inside computers as part of their operating systems. Tecnotes, the company Bonetti founded, gives discarded circuit boards a new life by turning them into long-lasting three-ring binders, address books, clipboards, notebooks, and memo books. The Sag Harbor, New York, firm also produces clocks, lamps, and key rings from the boards and is developing still more unusual and useful products.

Tecnotes pays a scrap fee to circuit-board producers, who reject 3 percent of all the boards they produce because of minor defects. Since its inception, the company has resourcefully converted more than fifty tons of circuit boards into objects of art that also serve as everyday items with a futuristic look. Mitch Davis, Tecnotes' president, says, "It's a difficult process to get rid of the waste created by these boards. It's not economical to melt them down, and it's expensive to haul them to landfills. Instead, we produce durable products that would be headed for the trash. Keeping these items out of landfills and creating useful products while using little energy is what we're all about. We've taken recycling a step further by reusing the material as it is—without any reprocessing."

Tecnotes products are sold in more than a dozen countries, including Japan, Australia, and Brazil, and sales continue to grow. Many of the products have been featured in museum exhibits displaying forms of art using computer or recycling themes, including a show at the Museum of Modern Art in New York City. The products can be found in gift shops, museum and environmental stores, and several mail-order catalogs. In addition, Tecnotes has expanded into the advertising specialty market, imprinting its products with the logos of companies ranging from IBM to community recycling organizations. "We also have gone on-line in a couple of 'cyber-malls,'" Davis says. "We see the Internet as a great potential distribution channel for us because technical people have always loved our stuff."

Because circuit-board configurations vary greatly, each Tecnotes product has a unique look. Most of the boards have a green background and are covered with varying patterns of copper lines that make them look like futuristic city maps. The binders are held together with hinges made from recycled plastic or steel. "Everyone feels good about recycling," Davis says. "Our products simply let them reutilize something at the same time."

In Texas, a different company is reusing yet another type of computer product. Each year, an estimated one billion computer disks are purchased in the United States. Stacked on top of one another, they would form a pile more than two thousand miles high. Surprisingly, millions of these disks never even touch consumers' hands before heading to a landfill. When a software company upgrades one of its products—going from, say, version 3.1 to 3.2 of a word-processing program—thousands or even millions of unused computer disks are suddenly rendered useless. This is because the labels pasted on them to identify the software product are difficult to remove without damaging the disks. In the past, these disks landed in the nearest landfill. Now a company named Eco Tech is taking the disks and giving them new life through a unique recycling idea. Eco Tech purchases disks containing obsolete versions of software and creates blank disks ready for use, says the company's general manager, Rick Wynn.

A friend of one of the company's original founders encountered the problem firsthand while working for a computer company. He watched as truckload after truckload of obsolete software disks were carted off to the landfill. The cost of reclaiming the disks, by removing the useless adhesive labels and erasing the programs on them, were too high to justify an in-house effort. Then, sitting at a kitchen table one night, Eco Tech's founders thought of a way to remove the outdated labels from commercial software disks—the main obstacle to recycling them. Through trial and error, they eventually developed a machine to handle the task. The company now contracts with major software and computer manufacturers to obtain 3.5-inch disks loaded with obsolete software.

When Eco Tech receives packages of software, all manuals and cardboard boxes are first separated out for recycling. The diskettes are then sorted into categories, such as high density or double density. Next, the disks are erased using an electromagnetic device that eliminates all data as well as any viruses. Any warped or damaged disks are broken down into parts, which also are eventually recycled. The labels on the remaining disks are removed through a proprietary process that uses no chemicals. After delabeling, they are formatted for use in either IBM-compatible or Macintosh computers and packaged for resale in recycled paper boxes. Each disk carries a lifetime warranty, and new labels printed on recycled paper with vegetable ink are provided. Because the manufacturers Eco Tech contracts with use only the highest quality materials, the disks are superior to many other versions on the market, Wynn said. Additionally, they have only been written to, or programmed, once, and never used. Called Eco Disks, the products are indistinguishable from any other blank disks used to store computer-generated information.

For a service fee, Eco Tech also provides internal recycling for large companies by taking existing disks, reformatting and delabeling them, and then returning them to the companies. Additionally, interested companies can send Eco Tech their obsolete disks and receive a rebate in reprocessed disks worth 20 percent of the total disks reclaimed. "The company has now processed upwards of 500,000 disks, and handles about 40,000 a month now," Wynn says. "Before, they were just thrown in the trash by the millions—and they are there forever. We're simply repackaging something that hasn't even been used before."

Although the thought of a few thousand—or even million—disks being tossed into landfills may not be particularly discouraging because they do not take lots of space, reusing the disks illustrates an important recycling idea. By putting more thought to what can be recycled in modern society, we can gradually trim our total waste stream. This will not only prolong the life of our landfills but will also curb the need to produce new products and materials from virgin sources. From small items such as computer disks to larger products such as monitors, hard drives, and keyboards, anything than can be reused adds to the cumulative benefits recycling provides the environment.

FOR MORE INFORMATION

Computer Recycling Center, 1245 Terra Bella Avenue, Mountain View, CA 94043; (415) 428-3700

Eco Tech, 11450 F.M. 1960 West, Suite 208, Houston, Texas 77065; (800) ECO-6175

Tecnotes, P.O. Box 3140, Sag Harbor, NY 11963; (800) 331-2006

Composting is another form of recycling. In addition to being found in engaging and varied forms, Zoo Doo is high-quality compost. (Photograph courtesy of Earth News. Used with permission.)

CHAPTER 2

THE DIRTY BUSINESS
OF COMPOSTING

A composting recipe book?

From dead chickens to old potatoes to crab scraps—if a smelly material is causing problems, chemist Will Brinton has a recipe to make it go away. As president and founder of Woods End Research Laboratory in Mount Vernon, Maine, he is developing new methods to compost animal and vegetable garbage that decomposes too slowly and smells too bad to be tossed into a normal compost pile. In doing so, Brinton is creating a new way for businesses to dispose of large quantities of undesirable material. Instead of paying high prices to have the offensive matter accepted at a landfill, they can have Brinton whip up a recipe to make the material decompose. The remaining compost is then ready for farm or garden use.

For example, when a fire smothered thousands of chickens owned by a Maine egg producer, the company buried the carcasses on its land. However, state environmental officials—fearing groundwater contamination—told the company to dig up the birds and dispose of them properly. With a thousand tons of rotting chickens on its hands, the egg company decided to call Woods End rather than pay the high cost of having the material hauled to a landfill. Brinton and his composting team at the laboratory first analyzed the birds for their chemical composition, then started adding various substances to the material to see which ones promoted rapid decomposition. Eventually, a mixture of carbon, sawdust, and chicken manure was used to turn the mess into compost in a matter of months, instead of years.

"At more than seventy dollars a ton to have waste dumped at landfills, in many cases it's more cost effective to have it composted," Brinton says. "We can take your waste material, analyze it, and tell you what kind of compost it can make. Between 12 and 30 percent of the waste hauled to landfills could be recycled through composting."

The plant and soil scientist uses a computer program he developed to determine the most effective way to compost different organic wastes. By finding the right mix of materials and combining them in proper proportions, Brinton says he can prompt almost any waste to quickly decompose into high-quality compost. "By diverting these wastes, we're buying time for our landfills. And compost is much better for fertilizing than chemicals because it helps rebuild the soil naturally."

For well over a decade, Woods End has researched the effects of compost on soil and plants using both laboratory and greenhouse testing. Through this work, the lab has found that improper composting practices can lead to unstable soils and may even be detrimental to plant growth. "The amount of compost being produced in this country is increasing at a rapid rate. However, it appears a substantial portion of this compost is being produced with little regard to technical merit and effect on plant growth," Brinton asserts. "As consumer awareness of compost quality begins to grow, future composting efforts will need to address the production of quality compost."

To that end, Woods End has created a compost-maturity kit for testing the stability of compost mixtures and their readiness for use in farming, gardening, or other applications. The easy-to-use, low-cost procedure allows for quick on-site decisions about compost maturity. A fresh compost sample is scooped into a calibrated jar, and a plastic paddle—complete with appropriate chemicals—is then placed in the sample. In four hours, a colored bar on each paddle, which measures carbon dioxide production and heat, reveals the maturity of the compost. Yellow on the color bar means raw compost, orange active compost, red advancing compost, and purple very stable compost.

To further promote the practice of composting, Woods End also has created a compostable food-scrap bag for household and commercial or institutional use. The bag provides a convenient and clean way to collect compostable food scraps, Brinton says. Called the Food Cycler, the 1.5-gallon moisture-resistant bag is made of recycled paper, sports a special cellulose liner, and has a reinforced square bottom so it will stand freely. The compostable bag can hold both wet and dry food scraps—including flowers, tea bags, cheese scraps, eggshells, paper napkins, paper towels, coffee grounds and filters—for up to a week. It can then be placed in the backyard compost bin, or at the curb, if a municipal composting program is in operation. Larger versions of the Food Cycler are available for use by the food service industry to promote composting by restaurants, cafeterias, dining halls, and grocery stores.

The product has been used successfully in several large-scale collection projects in Greenwich and Fairfield, Connecticut, Mississauga, Ontario, and

New York City. When combined with other traditional recycling practices, the use of the Food Cycler has reduced participating households' overall solid-waste stream by up to 50 percent, according to Brinton. The bag, with its unique breathable construction that eliminates waste odors, also provides savings in wastewater treatment by reducing the amount of food scraps washed into sewer systems through sinks and garbage disposals.

On another composting front, an odorless method of treating human waste without water or septic systems developed nearly fifty years ago in Sweden is gaining popularity in the United States. Composting or recycling toilet systems are being used in more and more parks and other outdoor settings to save both water and money, says Kevin Mart, a division president with Clivus Multrum Inc., based in Cambridge, Massachusetts. In Latin, *clivus* means "sloping or inclining," and *multrum* is Swedish for "composting room." The process has been described as a forest floor in a polyethylene tank.

The toilet systems emulate nature by using natural biological decomposition processes to convert human waste into small amounts of safe compost. Maintenance workers need only add minimal amounts of sawdust or wood shavings periodically as a bulking agent. With the help of the sawdust and air channels that provide oxygen—and without any chemicals or other additives—the waste materials slowly decompose over a matter of months. Because human waste is 95 percent water, only a small amount of compost is eventually produced. A Clivus toilet unit receiving 36,000 uses a year will produce a mere five cubic feet of compost, Mart said, which can be used just like any other form of compost, such as in gardens or flower beds.

An air-vent and blower system keeps bad odors at bay by pulling air down through the toilets themselves. And disease-causing organisms die because conditions produced in these composting toilet systems will not permit them to grow. They are consumed by active biological agents created in the process of decomposition.

Clivus Multrum has installed its composting toilets all over the United States, as well as in China, Korea, Cuba, Nicaragua, and other countries. The director of the U.S. National Park Service even installed one in his home. After the completion of Interstate 70 through Colorado's Glenwood Canyon, the canyon's five new rest areas all were equipped with Clivus Multrum toilets. The highway winds through the very narrow canyon, and the rest areas sit adjacent to the Colorado River. "In an environment like that, normal flush and septic systems aren't economical because of money and land requirements," Mart explains.

Colorado now has hundreds of the units installed at various parks and trailheads throughout the state. In Massachusetts, the state's Audubon Society plans to save more than 100,000 gallons of water a year through the use of composting toilets at its new headquarters building. By flushing wastes, modern society uses approximately eighty pounds of potable water to handle one to two pounds of waste, creating roughly eighty-two pounds of sewage. "It's a product that does not pollute," Mart says. "Compared to flush systems that send waste into the ground, it's a much cleaner system. We purify the wastes and produce an end product without using any water."

Human waste isn't the only form of animal waste that can be composted. Zoos around the United States are joining in an effort to compost the tons of manure created by their animals. On a visit to Singapore in 1990, entrepreneur Pierce Ledbetter saw elephant dung sold as compost at numerous garden shops and nurseries. He noticed that it fetched twice the price of other types of compost. Elephant manure, it turns out, has a high level of nitrogen due to the animal's diet and is therefore more beneficial as compost. Having been involved with the Composting Research Center at New York's Cornell University, Ledbetter soon developed an idea.

He approached the Memphis Zoo about the possibility of composting the zoo's animal manure and selling it. Zoo officials immediately raised a number of objections—too much time and effort to compost, too smelly, and no one would want the end result. "So I said, 'Let me address these issues,' and in the end, they agreed to give it a shot," Ledbetter recalls. "I agreed to take care of both the composting and marketing."

It took six months of composting to produce the first batch of benign-smelling Zoo Doo. When a Memphis newspaper published a story to announce the product, the first batch sold out in the first weekend. When radio personality Paul Harvey aired a segment about the product on his national news show, Zoo Doo's toll-free phone number began ringing and didn't stop. "People wanted to buy it for themselves, for their friends, or maybe for some in-laws they didn't like," Ledbetter remembers. "I was on the phone all day the first day, and the phone company later said more than sixteen thousand people tried to get through."

Ledbetter's initial idea revolved around people coming to the zoo and buying bushel baskets of the compost. Because thousands of baskets are used for feeding the animals and then discarded, this presented another way to recycle. But when the idea took off, Ledbetter soon began selling Zoo Doo through the mail in smaller one-, five-, and fifteen-pound buckets. A dozen zoos now participate in the Zoo Doo effort, including the famous San Diego Zoo. They all end up saving thousands of dollars each year simply by not hauling tons of manure to a landfill every day, and they also receive a large percentage of the profits made from selling Zoo Doo.

"The zoos didn't realize how much they were spending on disposing of these wastes," Ledbetter says. "An average zoo sells twelve thousand dollars worth of Zoo Doo a year with little effort, but ends up saving thousands more in landfill fees." Additionally, gardeners benefit from the high quality of the compost. Zoo Doo is produced from a highly controlled waste stream—no sick or meat-eating animals contribute to it. That's because what's better going in ends up being better coming out, Ledbetter explains.

"But awareness of composting is the best thing we've accomplished. Zoo waste is tiny compared to the amount of yard wastes that go to landfills every year. The zoos set up composting exhibits to show kids how easy it is to compost. And there's no better tool than an elephant's butt to create awareness and humor about composting. It's so much more exciting than a rotting pile of leaves."

Granted, composting isn't the most effortless task in the world. It does take some work and an ongoing effort to set up a composting bin and create usable compost. But it is relatively simple work, and as can be seen in this chapter, a significant portion of the waste stream, mainly in the form of food scraps and yard waste, can be recycled through composting.

If, however, the whole concept makes you uneasy, consider encouraging your restaurants and other food services to compost their wastes. These operations generate large amounts of garbage that can be composted. Or ask your local government to consider the idea of organizing a municipal composting effort that restaurants could participate in. Additionally, local and state governments could consider using composting toilets for parks, and zoos could join the Zoo Doo effort. By creating a growing awareness of the many benefits of composting in our society, we can continue to reduce our waste stream and improve the health of our soils.

FOR MORE INFORMATION

Clivus Multrum Inc., 104 Mt. Auburn Street, Cambridge, MA 02138; (800) 962-8447

Woods End Research Laboratory, P.O. Box 297, Mt. Vernon, ME 04352; (207) 293-2457

Zoo Doo Compost Company, 281 East Parkway N, Memphis, TN 38112; (800) I LUV DOO (458-8366)

Paper, paper, and more paper . . . how can we keep it from stuffing our landfills? The MAPelopes *product line is as wide open and horizon-expanding as the maps it recycles. (Photograph courtesy of Forest Saver, Inc. Used with permission.)*

CHAPTER 3

THE PAPER CHASE

Paper is one item modern society probably will never be able to do completely without. Some people argue that eventually the computer age will create a paperless society, but most computers today are attached to printers loaded with paper. Although cutting down on the use of paper in its many forms is a desirable challenge, totally eliminating its use probably will never be possible. However, by recycling paper and reusing the resource instead of returning time and again to the forests of the world for virgin supplies, paper use in society can become much less of an environmental issue.

Manufacturing paper from virgin materials, as opposed to producing it from recycled sources, affects the environment on many fronts. Logging forests to gather trees needed to make paper—in many cases through the practice of clear-cutting—has an obvious negative impact on the environment. Our forests are already disappearing rapidly as lumber is taken to construct furniture, houses, and numerous other products. And increased logging requires that additional roads be cut through shrinking forests. Although operations associated with logging provide jobs in our society, so too does the gathering of wastepaper and manufacturing of recycled paper.

The U.S. Department of Agriculture estimates that the average American uses the equivalent of one mature tree's worth of paper and wood products every year. Multiply that by upwards of 300 million Americans and the positive benefits of recycling paper products come into clearer view. The mountains of wood and paper waste hauled away to landfills every year have been dubbed the next great forest. The Environmental Protection Agency estimates that waste wood and paper account for more than 50 million tons of U.S. trash every year—nearly three times the 1990 timber harvest in the country's national forests.

Paper in all its forms accounts for 40 percent of all solid waste generated in the United States. In recent years, the recycling of newspapers has become popular—to the extent that many churches, community centers, and municipal recycling programs gather them for reprocessing. Unfortunately, office paper, which accounts for one-quarter of that 40 percent, still lags behind newspapers when it comes to recycling. However, as overall awareness of recycling in our society continues to grow, the collection of used office paper is also starting to make headway.

In San Francisco, a group of large businesses initiated a recycling effort centered on office paper. The Recycled Paper Coalition is a group of California companies hoping to stimulate more demand for paper products made from recycled materials simply by purchasing large quantities of them. Founded in 1991 by Pacific Gas and Electric Company, Bank of America, Pacific Bell, Safeway grocery stores, and three other smaller companies, the coalition now totals more than 160 businesses and nonprofit organizations.

"It all started when several large paper-using companies put their heads together and began asking questions about their paper use. They held meetings with representatives from each company's environmental department," says Marlene Meyer, executive director of the coalition. "Sometimes these large industrial companies really do have their hearts in the right place—it's been very heartening to see. Our coalition wants to encourage the use of recycled paper because increasing demand is crucial to developing a reliable, cost-effective supply of recycled paper. If enough major paper users convert to recycled paper, this will demonstrate long-term demand and send a signal to the recycled-paper industry to keep expanding and moving forward."

These businesses are to be commended for breaking with the more typical practice of buying paper made from virgin materials and then sending wastepaper on to the landfill. It costs money to set up and operate a recycling program as well as to purchase higher-priced supplies of recycled paper. However, as Meyer points out, once the recycled-paper industry has expanded to meet the increased demand created by such recycling efforts, cost differences will disappear and the environment will benefit.

Companies and organizations participating in the Recycled Paper Coalition pay no dues or annual fees. The only requirement is that each company's chief executive officer or president sign the coalition's charter that calls for members to:

- Reduce waste by using paper products more efficiently.
- Whenever possible, purchase recycled paper supplies made with a minimum of 10 percent postconsumer waste content.

- Work with paper manufacturers to increase postconsumer content as quickly as technology will permit.
- Start a comprehensive paper recycling program.
- Produce an annual plan outlining recycling goals for the upcoming year and an annual report detailing successes and failures.

As the San Francisco group began demonstrating promising results, two new satellite recycled-paper coalitions formed—one in California's state capital, Sacramento, and another in southern California. A year later, a number of companies in Dallas, Texas, joined to become another chapter of the coalition. And in 1995 Chicago joined the effort to make recycled-paper products the choice of major institutions throughout its metropolitan area. "In a few short years the coalition has grown from a local Bay Area group to a national organization seeing tremendous results," Meyer said.

In 1994 the coalition diverted more than eighty thousand tons of paper from landfills—a massive amount of paper that would cover the area of a football field to a depth of nearly 150 feet. Participating organizations purchased nearly 150,000 tons of recycled paper with an average postconsumer waste content of 26 percent. (Postconsumer waste is paper products that have actually been used by consumers and then recycled.) Coalition members also recycled a total of more than forty-three thousand tons of office paper during the year, and nearly all the participating organizations used recycled paper for their annual reports, newsletters, and training manuals. Additionally, in many cases members' purchases of paper towels and tissues were of products with 100 percent postconsumer content.

"Our membership continues to grow every day as more people hear about us, and we encourage more and more organizations to join us," Meyer says. "It is the supply-and-demand concept: The more members we have, the more positive impact we can have on the market and, ultimately, on the environment. Coalition members bring purchasing strength to the recycled paper market by demonstrating demand for its products. This sends a clear message to paper manufacturers and suppliers across the country—a message that the demand for recycled paper is on the rise, and the industry must take measures to meet that demand with price-competitive products."

In another effort—this one initiated and researched by the U.S. Department of Agriculture—a new type of construction material has been developed that is composed strictly of recycled wastepaper, cardboard, and other forms of trash that otherwise could not be reused. Called Spaceboard, the product uses wood-based fibers that are molded into panels and can be substituted for particleboard, plywood, and other nondurable construction materials. Scientists at the Forest Products Laboratory in Madison, Wisconsin, operated by the Department of Agriculture and the U.S. Forest Service, developed Spaceboard in the late 1980s and continue their research work on its processing today.

The government laboratory, in existence for more than eighty-five years, originally brought together various national research projects involving wood, explains Ted Laufenberg, a program manager at the lab. One of the lab's earliest successes, not long after the turn of the century, involved wood-preservative research that extended the life of railroad ties and greatly reduced timber demand for ties. This was just one of many technical developments that have helped the lab preserve forests. The lab's mission is to develop ideas that will conserve forest resources through better use of trees and all the products they are transformed into—including the final product, trash.

"The raw material in Spaceboard—cellulose fiber—is by far the most abundant waste material on earth," says Laufenberg. "We eventually could have the potential to divert 40 to 50 percent of the waste stream headed for landfills and recycle it into a useful product. The future environmental effects of this technology are just beginning to be comprehended. As with most recycling technologies, we may be witnessing an unfolding environmental and material technology revolution."

Gridcore Systems International, a company based in Long Beach, California, received the rights from the U.S. government to market the Spaceboard process in the construction and furniture industries and now operates a pilot processing facility. Gridcore, the company's product name for Spaceboard, currently is being used in the stage- and set-design industry in Hollywood, says Robert Noble, chief executive officer of Gridcore Systems. However, a portion of the entertainment industry still constructs sets for movies, television shows, and music videos with a type of plywood called "lauan" that is made from a tropical timber. Unfortunately, the lumber needed to make lauan is being logged heavily in the threatened rain forests of Southeast Asia, and alternatives to this lightweight and inexpensive wood are few in number. By switching to Gridcore, the entertainment industry can escape the need to purchase this rain-forest-destroying lumber.

Gridcore features a ribbed core that is covered by panels. It is formed when wet fiber pulps are spread across a mold and shaped into panels of a construction resembling a honeycomb. The thickness of the honeycomb cores varies depending on the intended use of the panel. Pound for pound, Gridcore is twice as strong as plywood and can weigh as little as a third of the weight of typical particleboard or plywood. And because Gridcore is made of 100 percent nontoxic raw materials, the product is itself recyclable.

Gridcore Systems' original interest in the material was for use in structural wall, floor, and roof systems for low-income housing, Noble says. With the number of new homes built in America every year, construction materials made from waste products could prove an exemplary form of recycling. The company now is focusing on markets for shelves, cabinets, furniture panels, and wall-partition systems such as those used for the various exhibits prevalent at trade shows made of Gridcore. In addition, the Department of Agriculture has licensed the technology to a South Carolina company for use in packaging and to another California company for use in transport vehicles. Other possible uses for Gridcore include skis, book covers, suitcases, surfboards, doghouses, and even the walls of railroad freight cars.

"Spaceboard continues to generate a lot of interest in many different industries, and the government is looking at more sublicensing to promote its use," Ted Laufenberg says. "It's a material that can accommodate waste fibers that are contaminated with ink, dirt, plastic, and adhesives—material that doesn't meet the quality standards needed for other recycled-paper products. And it's an excellent alternative for many nonstructural building materials."

Although recycling paper products is a fine and admirable practice to help the environment, dodging the recycling process altogether and simply reusing paper products goes a step further. To that end, postal patrons now can send letters and other correspondence in style with MAPelopes—colorful and exotic envelopes made from surplus and outdated maps. Enzo Magnozzi, founder of Forest Saver Inc., which deals in recycled-paper stationery products, came across the idea by accident. While placing an order to have envelopes made, he threw in some old maps a friend had given him. Magnozzi wanted the white side of the maps on the outside of the envelopes, but the envelope maker misunderstood and did the opposite.

"And people just loved the product," he says. "It evolved out of a mistake and is now our major product. We're selling five tons worth of old maps every month, and these products are better than recycled paper because they are being reused instead of recycled. It takes energy to produce recycled paper. The maps are ready to go—it's simply second use."

The map idea has evolved into other product lines as well, such as writing sets and notepads. Some people like writing on the map side if it is faint enough, while others prefer the white or backside of the products. And map lovers enjoy trying to figure out what part of the world a MAPelope represents. The new-wave envelopes spare trees and landfills as well as the chemicals and water needed to produce recycled paper.

Forest Saver's main supplier of maps has been the U.S. government, which initially provided a free supply of outdated topographical maps. But as interest and sales have grown, other sources have been tapped, such as the Rand McNally Company and other mapmakers. Because of the success of the idea, however, the company now must purchase the maps. Forest Saver's MAPelopes alone divert sixty tons of paper from landfills every year, and maps turned into stationery and notepads add to that amount.

MAPelopes and other Forest Saver products, which now include a complementary line of recycled wrapping papers called Mapwrap, are sold in environmental products stores, museum shops, college bookstores, and some large retail chains around the country as well as in various mail-order catalogs.

Magnozzi's personal favorites are MAPelopes made from maps of coastal areas with cities interspersed. He's also fond of the Great Salt Lake and says he enjoyed making products out of obsolete world maps after the Soviet Union disintegrated several years ago. "But there are so many nice ones, it's hard to say. The color schemes on different maps can be great. The products are not only intrinsically interesting, but they are better for the environment because one day the material was an old map, and the next it's something else."

Ideally, more and more types of paper will be diverted from landfills as the recycled-paper industry continues to expand. New types of labels and cellophane windows are now on the market, allowing more and more paper products to be recycled. And much of the newspaper industry relies on recycled newsprint. Average citizens can help the recycled-paper effort by collecting office paper and recyclable junk mail at home, then either hauling the paper to the office to be recycled or to a recycling center. And remember to recycle your newspapers. The more paper trash we can cut from our overall waste stream, the more trees will be spared and the longer our landfills will have room to spare.

FOR MORE INFORMATION

Gridcore Systems International, 1400 Canal Avenue, Long Beach, CA 90813;
(310) 901-1492

MAPelopes/Forest Saver, 1860 Pond Road, Ronkonkoma, NY 11779;
(800) 777-9886

Recycled Paper Coalition, 3921 East Bayshore Road, Palo Alto, CA 94303;
(415) 985-5568

Plastic milk jugs and soda bottles hold more than your favorite beverages. Gene Pendery, left, and Bob Williams of Recycled Plastic Products, Inc., display some of the plastic milk jugs that will be used to manufacture Plasti-Fence. (Photograph courtesy of Recycled Plastic Products, Inc. Used with permission.)

CHAPTER 4

RECYCLED PLASTIC FINDS A HOME

Innovative uses for recycled plastic are multiplying and gaining in popularity. These new uses provide environmental benefits by allowing old plastic to take the place of other virgin materials, such as wood and concrete, and by keeping it from piling up in landfills. To keep the recycled-plastic industry growing, however, the current collection system for plastic bottles and containers must be improved. Plastic recycling is a newer and more difficult process to execute than are other forms of recycling for materials like glass or aluminum. Plastic is produced in a number of different types, which must be separated before recycling can take place. Nonetheless, increased supplies of recycled plastic will spur more demand and create a broader market for the material.

On one front, recycled plastic recently expanded into the multi-million-dollar world of fashion design, and the development is good news for the environment. Wellman, Inc., one of the world's largest polyester and nylon fiber producers, is also the world's largest and most advanced recycler of plastic. In 1993 the company launched Fortrel EcoSpun—a revolutionary fiber made from 100 percent recycled plastic PET (polyethylene terephthalate) soda, water, and food containers, says Jim Casey, president of Wellman's fiber division. "What makes EcoSpun such an exciting breakthrough is that it gives consumers many more options to close the loop in PET recycling. EcoSpun's success is proof it can be fashionable to be environmentally proactive."

In its earliest incarnation, the recycled fabric was used for rugged outdoor apparel because it lacked a fine texture. Wellman's ongoing research and development, however, has led to a softer feel and thus broader applications of the material, including use in socks, sweaters, backpacks, luggage, golf shirts, and fleece jackets. "For every pound of 100 percent EcoSpun, approximately ten PET bottles are kept out of America's landfills," Casey says. "We reclaim up to 2.4 billion bottles every year." Companies like Patagonia, L. L. Bean, and Lands' End already have featured EcoSpun products in their clothing lines. Nordstrom,

Saks, Macy's, and J. C. Penney also have distributed various products made from the material. Wellman has had so much success with EcoSpun that the company is operating at capacity. "When we started in April 1993, we only had one manufacturer using the fabric. Now, well over one hundred manufacturers are using EcoSpun in everything from thermal underwear to eco-denim in jeans," Casey says.

EcoSpun is made from containers used primarily for food, and these bottles are rigorously controlled by the U.S. Food and Drug Administration and are made of exceptionally high-quality plastic. When cleaned, melted down, and drawn out, they produce an excellent fiber. After EcoSpun fiber is spun into yarn, which is knitted and woven into fabrics, it becomes extremely versatile, making everything from velvet upholstery to linenlike cloth to thick pile material for jackets. It can be comfortable and breathable for use in clothing, or canvaslike and waterproof for luggage.

"EcoSpun is strong, washable, dryable, and holds color," Casey says. "Right now, you won't find a real silky blouse made of EcoSpun, but we're working on it. We're trying to make it even finer, which will make the fiber even more versatile. Every day, fashion professionals make thousands of decisions that have an impact on this fragile planet. EcoSpun is a new fabric that can help them conserve the world's resources, whether as 100 percent polyester or in blends with cotton, wool, nylon, and other fibers."

In 1992, Wellman recycled 100 million pounds of plastic beverage bottles, which contained the equivalent of enough fuel oil to power a city the size of Atlanta for one year. (To produce virgin plastic, petroleum is needed, so any plastic recycling effort helps reduce the industry's consumption of oil.) "Unfortunately, there's a bottle shortage," Casey explains. "The price of recycled bottles has doubled in the last year, and future growth in this area is dependent on a national bottle recycling law. Right now, only eight states have mandatory recycling laws for PET bottles, and there hasn't been a new one added since 1985."

Despite the shortage, PET bottles already represent a recycling success story. Nearly 40 percent of all PET bottles in the United States are recycled; compare that to the 3 percent figure for all plastic. "We're closing the loop by giving people back a Patagonia jacket or a backpack," Casey says. "People are saying, 'There's a payoff here.'"

Wellman began business in 1927 and is based in Bridgeport, New Jersey. The Fortune 500 company started making nylon and polyester fibers from recycled raw materials in the mid-1960s at its plant in Johnsonville, South Carolina. In 1972, Wellman International began operations in Ireland to produce similar polyester and nylon fibers for various European markets. To this day, the Irish plant and Wellman's two South Carolina plants manufacture materials exclusively from 100 percent recycled plastic. Overall, 40 percent of the company's total fiber production comes from recycled materials.

Another company is using 100 percent recycled plastic to build woodlike fences and decks. Calling its product "the last fence you'll ever build," Recycled Plastic Products, Inc. is turning plastic milk bottles into long-lasting fences. The recycled material for the fencing is bought after the bottles have been ground into pellet form by various plastic recyclers. Any drink or bleach bottle made with HDPE (high-density polyethylene) plastic can be used in the fencing products, explains Gene Pendery, president and founder of the Denver, Colorado, company. Pendery, a geologist, decided several years ago to develop an idea he had with his partner, Walter Perkins, that used recycled materials. The product they eventually created is called Plasti-Fence.

Initial production of the fencing started in late 1992, and a typical six-foot-high privacy fence was test-marketed in the Denver area through a chain of home-improvement stores. The product sold well, in part because Plasti-Fence is priced competitively with premium wood fencing material. Pendery's plastic version is less expensive than either pine or redwood but is about 40 percent more expensive than cedar fencing. Plasti-Fence's patented shape helps keep construction costs down, and when the lower maintenance costs are factored in, plastic fencing is even more beneficial.

The fence is produced with a simulated wood grain in brown, country white, and a cedarlike color and closely resembles painted wood. Plasti-Fence is installed with the ordinary tools used for traditional wood fencing. It can be sawed, screwed, or bolted like wood. The product comes with a twenty-year limited warranty—better than any wood products, Pendery asserts—and it should easily last more than twenty years because it doesn't rot or splinter. In addition, Plasti-Fence is produced with antioxidants and ultraviolet-light protection to reduce fading and eliminate the need for painting. In fact, the fence is virtually graffiti-proof because paint won't permanently adhere to HDPE plastic.

"A one-hundred-foot section of our fence keeps five thousand one-gallon plastic containers out of landfills," Pendery explains. "And no trees need to be cut down to make it." Plasti-Fence comes with recycled plastic posts and rails, as well as pickets, to make a complete fencing package. The entire line of Plasti-Fence products has been approved by Sears for its home-service line of products. Plasti-Fence also has expanded to markets in Texas, California, Illinois, and Georgia, and the product is sold on a wholesale basis to general contractors. In addition, the company is introducing maintenance-free plastic lumber for decking and recycled-plastic tire stops for parking lots.

In the construction industry, one company has found yet another unique use for recycled plastic. ThermaLock Products developed a new building block partially composed of recycled plastic that can take the place of normal concrete blocks in construction projects to create energy-efficient buildings. Although the ThermaLock Block is similar to the concrete blocks used in many buildings, this new product consists of a thick layer of molded plastic sandwiched vertically between two layers of concrete. When the blocks are cemented in place, the resulting wall takes shape with an extra layer of permanent insulation inside. The block's plastic layer contains up to 30 percent recycled plastic and is designed to make walls five times more resistant to heat loss than other building blocks, according to Ken Blake, one of the company's cofounders. "The primary purpose of the ThermaLock Block is to save energy," he says. "But we're also showing how we can bury plastic waste in walls instead of our landfills."

Current building codes limit the amount of recycled plastic that can be used in building construction to 30 percent because of concerns about strength and existing standards, but Blake hopes U.S. regulations soon allow the use of 50 percent recycled plastic. In Europe, 100 percent recycled plastic is allowed in similar applications. About five billion concrete blocks are produced in the United States every year. ThermaLock's immediate goal is to capture just 2 percent of this market.

A typical concrete block has a web-shaped interior design that connects one half of the block to the other. The ThermaLock Block replaces this concrete webbing with the layer of plastic. "Our block is the only one on the market without webs, which unfortunately transfer heat or cold directly from one side of the wall to the other," Blake says. "There are no thermal breaks in the ThermaLock wall—it gives superior insulation value to a building." The product is currently available in Florida, Illinois, and the Northeast, but the company is establishing new distribution networks in other parts of the country. "We're looking for partners that are already in the block business," Blake explains.

Thanks to efforts like these, markets are expanding for recycled plastic. There is no reason recycled plastic can't continue to replace other precious natural resources, such as wood, in many different applications. However, if society wants to reap the environmental benefits provided by plastic recycling, more people must support the system both by joining in the overall recycling effort and by purchasing products made with recycled plastic.

FOR MORE INFORMATION

Plasti-Fence, 2331 West Hampden Avenue, Suite 148, Englewood, CO 80110; (800) 235-7940

ThermaLock, Inc., 162 Sweeney Street, North Tonawanda, NY 14120-5908

Wellman, Inc., 1133 Avenue of the Americas, New York, NY 10036; (212) 642-0793

For a free guidebook to community PET plastic bottle recycling, write the National Association for Plastic Container Recovery, 3770 Nations Bank Corporate Center, Department P, 100 North Tyron Street, Charlotte, NC 28202.

Oil spills can devastate aquatic plant and animal life, the effects of which can last for decades. Sea Sweep is used around the world to clean up oil spills of all sizes—and can then be recycled as a petroleum product. Here, William Mobek prepares a demonstration in Indonesia. (Photograph courtesy of Sea Sweep, Inc. Used with permission.)

CHAPTER 5

REMEDIES FOR OIL SPILLS AND THEIR ILLS

When the *Exxon Valdez* spilled its vast cargo of oil into Alaska's fragile shoreline ecosystem, no topical agent was available to help minimize the damage the oil would cause as it spread. And with about ten thousand spills occurring on America's waterways every year—one spill every fifty-three minutes—there is plenty of damage to try to minimize.

Mammoth oil spills in oceans and other large bodies of water affect life at many levels. In Alaska, for instance, the surface and coastal contamination killed animals like otters, as well as numerous species of birds whose feathers become saturated with oil and unable to protect them against cold water or to allow them to fly. These dead animals were eaten by grizzly bears and eagles, which were then poisoned themselves. In addition, oil that flows under water kills a portion of the fish population, as well as the plankton that is lower on the food chain. Some of the oil sinks to the bottom of the ocean floor and mixes with sediment. Contamination at this level affects bottom-dwelling creatures such as lobsters and crabs. The cumulative results of an oil spill can obviously be devastating to numerous species and can last for decades.

But now a product made of a sawdustlike material helps stop the destruction caused by oil spills by gathering the oil in a nontoxic manner. Called Sea Sweep, the product is unique because it absorbs oil while repelling water. It is produced through a patented heating process that actually changes the structure of wood particles, explains William Mobeck, cofounder of Sea Sweep, Inc., located in Denver, Colorado.

Mobeck's partner, Dr. Thomas Reed, a retired research professor from the Colorado School of Mines, developed the process using no chemicals or additives—just heat. The resulting product is very effective. "It works like a sponge, soaking up oil in less than a minute, then floating indefinitely until it can be picked up and properly disposed of," Mobeck says. "And Sea Sweep won't contaminate shorelines if it washes up on beaches. It holds on to the oil until it can be collected."

In 1993 Sea Sweep was honored as one of the most technologically significant new products of the year by *R&D Magazine*—a periodical that monitors and publicizes technological developments of all kinds. That same year, Sea Sweep also earned a gold medal at the Clean Seas International Conference for the company's "praiseworthy efforts in conjunction with the preservation of a clean marine environment." The product is now being used in countries around the world, including Canada, Chile, Colombia, Cyprus, Japan, Indonesia, Peru, Russia, and the United States. The U.S. General Services Administration includes Sea Sweep as a product that may be purchased by all agencies of the federal government, and the Environmental Protection Agency has authorized on-scene coordinators to use Sea Sweep in an emergency response to oil spills on navigable waters.

Sea Sweep can be applied directly to oil spills by any type of dropping mechanism, including a helicopter. It is then picked up with booms or skimming devices. The product can also be directly applied to oil spills on land, according to Mobeck, then swept up with a broom minutes later, after it has absorbed the oil. And Sea Sweep can be used on animals to help remove oil from fur and feathers. Once collected, oil-saturated Sea Sweep doesn't have to end up in landfills, either. The used product can be recycled by being burned in power plants or industrial furnaces—providing a use for what would otherwise be wasted petroleum.

The product has many potential users, including major oil companies, fire departments, hazardous-materials safety teams, spill-recovery units, boat marinas, airports, trucking companies, and even homeowners. Sales of Sea Sweep have been growing as word about the product reaches more potential users. The product is sold in five-, ten-, and twenty-five-pound bags, as well as in five hundred-pound containers for use in helicopters or other large vehicles. Sea Sweep must be bought directly from the company at the present time, but a distribution network for it will be developed soon. In addition, Sea Sweep, Inc. sells kits manufactured expressly for ships and fire departments, with everything needed to clean up a spill packaged inside a recovery barrel.

Each ton of Sea Sweep can absorb more than one thousand gallons of oil—about half a gallon of oil per pound. "No one realizes how many oil spills occur daily. I get a fax or phone call every day from someone dealing with a spill," Mobeck says. "If we can help with even a small percentage of these spills, we'll be doing a tremendous amount of good for the environment."

A company in Texas has developed a slightly different approach to dealing with oil spills. Nature's Way, a new family of products featuring literally trillions of microscopic oil-eating microbes is being used around the country to eat up spilled oil, automotive fluids, and other petroleum-based products—even on a household basis. The tiny microbes quickly turn petroleum products into water and carbon dioxide, says Gene Kaiser, chairman of NW Technologies, the Houston-based company that manufactures Nature's Way. "Every time a thunderstorm occurs, it acts like a giant toilet, flushing all kinds of motor oil and other petroleum products down into our sewer systems," he continues. "With Nature's Way, we've been able to create a biologically safe end-product from spilled oil, rather than have it contaminate our water supply." By taking care of thousands of small-scale oil spills before storms wash them into sewer systems, a product like Nature's Way could help protect water sources around the world.

Nature's Way is a two-ingredient product. Microbes are stored in a clay powder inside a small plastic bag. A nontoxic liquid that allows the microbes to flourish is the second ingredient. When the two are mixed together, the resulting solution is designed to work on dirt, asphalt, concrete, contaminated water, and other surfaces. The microbes first surround and devour the petroleum on top of the surface, then penetrate the area more thoroughly and feed on the oily substances until they are gone. In some cases the microbe population doubles every half-hour—accelerating nature's ability to cleanse itself. "Microbes have been around for billions of years, but they weren't studied much until the last fifty years," Kaiser says. "About twenty years ago, people began working with microbes to break down organic wastes. After that, researchers began searching for the strain that worked best on hydrocarbons."

The first person to develop microbes for breaking down oil and other petroleum compounds was Dr. Carl Oppenheimer at the University of Texas in the early 1980s. He discovered that certain microbes excrete an enzyme that breaks oil down and makes it easier for them to consume. After eating the oil, the microbes convert it into the energy they need to keep reproducing, Kaiser explained. As long as there is oil to be consumed, the microbes continue multiplying, which enables them to neutralize large quantities of oil in four to six weeks. Once the oil is gone, the microbes go dormant and depending on the environment can stay dormant for days or years without dying.

All Nature's Way products will work on emergency oil spills, but individual products are specialized for different types of applications. The product sells for ten to twenty dollars per gallon, and a gallon will treat an area of about two hundred square feet. An unopened bottle has a shelf life of about five years. Nature's Way has inaugurated the first widespread distribution of a certain kind of product to both consumers and businesses, Kaiser says. The product constitutes a form of *bioremediation*—using living organisms to devour waste products.

Steel mills, refineries, municipalities, automobile dealerships, and mechanics are popular customers. Many auto dealers in the Houston area use a fifty-five-gallon barrel of Nature's Way every month, for instance. As an added bonus, these businesses are finding that the microbes, when washed into holding tanks used to store oily wastewater—are also cleaning up the tanks and returning clean water to the environment, saving the companies from having to haul the wastewater to a dump. Individuals can use Nature's Way to remove oil from driveways, sidewalks, and garage floors.

NW Technologies is now working on a new Nature's Way product for consumers who change the oil in their cars on their own. Kaiser says the new product would allow do-it-yourselfers to simply add a mixture of microbes to a container of used oil and water, shake it, and return later to a biodegradable mass that could then be disposed of without worry.

A different type of oil-related invention has been created by a Wyoming businessman who hopes to clean up the oil industry by converting unusable wastes into usable products. Oil sludge is one of the world's greatest waste-management problems, says Neal Miller, president and cofounder of Centech, Inc. "American oil producers alone generate about 100 million gallons of the tarlike waste by-products each year," he said. Miller's invention can extract usable oil from the sludge in a safe and cost-effective manner. Called a three-phase centrifuge, the device employs the laws of gravity to separate the oil, water, and cakelike sediments out of the sludge. The centrifuge is akin to a giant washing machine during its spin cycle. As a washing machine spins rapidly, the clothes inside are forced to the outer edge of the drum, which separates most of the water from them. The Centech centrifuge operates in a similar fashion. Sludge is fed into the machine, then it spins at a high speed. Centrifugal force in conjunction with the patented design of the machine separates the oil, water, and sediment. Each component is then pumped through separate outlets for collection.

Chemicals are rarely used by Centech to help break down the sludge, unlike some previous treatments. The centrifuge does, however, incorporate a heating system that makes the sludge flow better. "Sometimes the sludge is so thick, it would almost be easier to shovel it into the machine than pump it," Miller says. "People all over the world have tried to build a centrifuge that would actually work on this scale, but they usually make a bigger mess than they started with."

For many years, the oil industry poured most of its sludge into open pits or buried it. Tougher environmental regulations in most states now have the industry searching for new disposal alternatives, not to mention ways to clean up the existing sludge pits.

Miller started Centech in 1986 in Casper, Wyoming. With just a high school diploma and plenty of determination, he succeeded where several engineers and multinational corporations failed. While working in oil field drilling operations, he saw companies try to use a more primitive two-phase centrifuge to separate oil from the sludge, but their efforts always failed. The idea fascinated Miller, and he became inspired to succeed where others failed. He went to Germany to observe a three-phase centrifuge and proceeded to build a similar, patented, version for oil applications. The technology has already earned international recognition. "I did all of my research and development work in the field," he says. "I have offers from Russia, Kuwait, and Latin America to use the technology."

Before Centech tackles the global market, though, the company wants to finish redesigning its product. "We're working with the Los Alamos National Laboratory in New Mexico to add some computerization to the control system," Miller says. "We're also making some adjustments to improve operating capacity. The next centrifuge built will have some drastic design changes to make it even more efficient. It will be larger, computer-controlled, and made of stainless steel."

The company started with only one centrifuge and had to turn down work because of its limited capacity. The centrifuge is mounted on a flatbed trailer and towed from one project to another by truck. The company's biggest job to date involved cleaning up a large active Wyoming oil field owned by Amoco. The six-month effort involved dealing with sludge pits, solvents, and oil-laden water. After treating more than 300,000 barrels of contaminated fluids, the centrifuge recovered more than 25,000 barrels of oil. The water recovered was reused in the oil field's operations.

Centech's next goal is to boost the size of its fleet to include a total of four centrifuges. The company also wants to test its device on other types of cleanups to see if the technology can be adapted to handle used motor oil, antifreeze, and even oil spills. Miller is willing to test his centrifuge on separating any mixture that isn't radioactive. "If you can contain the spill and pump it to the machine, I can process it," he says.

In a large-scale oil-recycling effort, a Texas-based company is taking used motor oil and giving it a second life. An estimated 1.3 billion gallons of used lubricating oil must be disposed of every year in the United States. Unfortunately, more than 250 million gallons of this oil ends up dumped down sewers, where it can harm the water table, out of carelessness. Another 177 million gallons are hauled to landfills. Fortunately, the nation's eighth largest refiner is now recycling used motor oil into gasoline and other petroleum products, becoming the first U.S. refinery to do so on a commercial basis.

As of 1995, Lyondell-CITGO Refining Company (LCR), of Houston, Texas, had processed 10 million gallons of used oil since it pioneered the technology used in its recycling program in 1992, says LCR's David Harpole. The company feeds the used oil into a heating unit. Operating at high temperatures, this unit breaks down the sludgelike waste oil into gasoline, heating oil, and petroleum coke, a solid hydrocarbon similar to coal. Eventually, LCR plans to process 30 million gallons of used oil a year, the equivalent of more than 25 million automotive oil changes. The company purchases the oil from two main sources, industrial companies that use large volumes of lubricating oil and used-oil collectors such as service stations. Although a majority of the oil now comes from Texas, Harpole said several barge loads—each containing more than 100,000 barrels of used oil—have been purchased from East Coast sources.

"What we have developed is a process that takes what was once a waste and converts it into a new product that has a strong demand—gasoline," Harpole says. Before LCR started processing the used oil, it had to demonstrate to the Texas Water Commission that no adverse environmental effects would result. Eventually the company was given the go-ahead from the commission and now operates under state regulations. "We're taking a waste material, ensuring it is not improperly disposed of, and providing an incentive for other companies to collect it and ultimately sell it to us," Harpole says. "This way we eliminate a potential contaminant and provide a way to reuse the oil."

As long as modern society must rely on oil and other petroleum products for any number of uses, oil will continue to contaminate numerous ecosystems around the world. Oil-well operations will continue to produce sludge, transporters will continue to have accidents, and users will continue to have dirty oil to discard. Given these factors, Mother Nature is fortunate that concerned entrepreneurs and scientists are trying to solve the various problems endemic to these activities. But there's plenty of room for other good ideas, too.

FOR MORE INFORMATION

Centech, 920 Lakeview Lane, Casper, WY 82604; (307) 265-7621

Lyondell-CITGO Refining Company, One Houston Center, 1221 McKinney, Suite 1600, Houston, TX 77010; (713) 652-4125

Nature's Way/NW Technologies, 5817 Centralcrest, Houston, TX 77092; (713) 680-1234

Sea Sweep, 2121 South Oneida, Suite 635, Denver, CO 80224; (303) 759-8118

PEOPLE POWER

Earth Day 2001?

Kids Face the Future

Helping Cultures Survive

Building Political Power for the Environment

Uniting Conservation and Tourism

Student Concerns

Environmental Education and Environmental Jobs

Earth Day events celebrate the power of collective concern for our planet and the life it sustains. (Photograph by Ebet Roberts, © 1995. Used with permission.)

CHAPTER 6

EARTH DAY 2001?

We've come a long way since that first Earth Day on April 22, 1970.

Or have we?

In many cases, the answer is a resounding *yes*. The practice of recycling is now commonplace, and polls show that most Americans consider themselves environmentalists, or are at least mindful of the need to protect the environment. However, in a number of other instances, the environment remains at risk today due to numerous forces that have kept its future in jeopardy ever since the Industrial Revolution created factories, pollution, and urban society. In addition, the populations of many of the world's countries continue to expand. This burgeoning mass of humanity breeds more poverty, as it increases demands on finite natural resources.

In the late 1960s, mounting environmental problems—including severe air pollution in a number of major U.S. cities and toxic waste pouring into the Great Lakes and other bodies of water—had raised concerns in the American public. Perhaps the Cuyahoga River catching fire piqued this cumulative concern. At the time, general opinion centered on the need to do something to halt this obvious destruction of the environment.

The overall goal of the first Earth Day revolved around creating a mechanism by which to express this environmental concern at national, state, and local levels. It aimed at inspiring the American people to demand change and action. To that end, it was a sterling success. At least in part because of the events of the first Earth Day, the U.S. Congress had within a year passed the Clean Air Act, the Clean Water Act, and also had created an environmental watchdog, the Environmental Protection Agency.

Earth Day 1970 united 20 million Americans in an effort to begin cleaning up a polluted environment. It was a bold first step. Along a stretch of West Virginia highway, one group of concerned individuals picked up five tons of trash, then dumped it on the steps of a nearby county courthouse to make a statement regarding America's lack of concern about littering and trash. Farther west, in Tacoma, Washington, a group of students rode horses down an interstate highway to call attention to society's overreliance on automobiles. The National Education Association estimated that 10 million public school children took part in various environmental activities such as tree plantings and essay contests. Gaylord Nelson, the senator from Wisconsin who created the idea of Earth Day, had seen his dream realized.

"For years prior to Earth Day, it had been troubling to me that the critical matter of the state of our environment was simply a nonissue in the politics of our country," Nelson once explained in a speech. "The president, the Congress, the economic power structure of the nation, and the press paid almost no attention to this issue, which is of such staggering import to our future. It was clear that until we somehow got this matter into the political arena—until it became part of the national political dialogue—not much would ever be achieved. The puzzling challenge was to think up some dramatic event that would focus national attention on the environment. Finally, in 1962, an idea occurred to me that was, I thought, a virtual cinch to get the environment into the political limelight once and for all."

Nelson persuaded then-president John F. Kennedy to undertake a national conservation tour designed to draw attention to the deteriorating condition of the country's environment and to propose ways to address the issue. In the fall of 1963, Kennedy began the conservation tour. Nelson and three other senators joined him on stops in Pennsylvania, Wisconsin, and Minnesota. However, his first idea did not work. The tour did not put the environment on the political map as Nelson had hoped, but he refused to let his hope of protecting the environment dim.

"While the president's tour was a disappointment, I continued to hope for some idea that would thrust the environment into the political mainstream," Nelson said. "Six years would pass before the idea for Earth Day occurred to me in late July 1969, while on a conservation speaking tour out West. At the time, there was a great deal of turmoil on college campuses over the Vietnam War. Protests, called anti-war teach-ins, were being widely held on campuses across the nation. On a flight from Santa Barbara to the University of California at Berkeley, I read an article on the teach-ins, and it suddenly occurred to me—why not have a nationwide teach-in on the environment? That was the origin of Earth Day."

Upon returning to Washington, D.C., Nelson sent letters to all of the country's governors and the mayors of major cities explaining the event and asking for support. He also sent a press release to all college newspapers. In a speech during the fall of 1969, Nelson announced the creation of Earth Day, saying the event would be held the following spring. The media pounced on the story, in part because of all the negative environmental news being covered by the press. "The response was dramatic. It took off like gangbusters. By December, the movement had expanded so rapidly that it became necessary to open an office in downtown Washington," Nelson said. "It was truly an astonishing grassroots explosion. The people cared, and Earth Day became the first opportunity they ever had to join in a nationwide demonstration to send a big message to the politicians—a message to tell them to wake up and do something."

Although momentum had been created, during that very first Earth Day celebration, Nelson uttered a prophetic question: ". . . Are we willing to make the commitment for a sustained national drive to solve our environmental problems?" Although environmental action continued and local Earth Day events were held in the ensuing years, it took twenty years to organize another one on a national scale.

The 1990 event marked a new beginning for Earth Day. Mainly due to the efforts of ambitious and well-known environmental activist Denis Hayes, whom Nelson had hired to organize the first event, its impact was felt around the globe as millions of people marked the start of a renewed effort. With more than 200 million people in more than 140 countries participating in various events, it became the largest peacetime event in the history of the world, according to 1990 Earth Day organizers. From a five-hundred-mile human chain in France suggesting humanity's need to work together to solve environmental problems to a group of five thousand Italians lying down on a highway to protest automobile exhaust, 1990's Earth Day took on a grand and global feel. Nelson, now a counselor for The Wilderness Society, said the 1990 version served a special purpose as an anniversary. "It gave people another chance to demonstrate their interest in environmental issues. And they made it clear they want these issues addressed."

For political reasons involving "turf" issues among some of the nation's large Washington-based environmental groups, Hayes had agreed to shut down the temporary nonprofit organizing body, Earth Day 1990, immediately following the event. However, after the twentieth anniversary, a new organization came to life to ensure that Earth Day activities would not experience another hiatus. The new nonprofit effort, called Earth Day USA, carried the 1990 event's momentum forward. Bruce Anderson, who had spearheaded New Hampshire's

Earth Day efforts in 1990, cofounded the organization with Nelson to help Earth Day remain an annual national occurrence.

"We had two goals in mind for the new organization," Anderson says. "We wanted to inspire people to get involved in Earth Day activities and provide them with needed services. Creating Earth Day USA basically told the world that this is an annual event and let interested people know that they are not alone. Secondly, we wanted to promote the event to the press and media, and started printing calendars that listed hundreds of groups' Earth Day–related activities that they could cover."

Earth Day USA now serves as the national coordinating headquarters and clearinghouse for Earth Day activities and information, serving millions of people in thousands of communities and organizations. It provides numerous contacts and a myriad of materials to groups and communities that want to organize Earth Day programs or events. The group also creates strategic alliances with various groups to further the impact of Earth Day.

"Earth Day's role has changed over the years. For the most part, its role in 1970 was to wake people up and let them know there was such a thing as the environment," Anderson says. "By 1990, everyone knew about the environment and that there were problems to deal with—so the focus shifted to individuals and their actions. There was a realization that the environment belongs to everyone and is everyone's responsibility. Today people are starting to target Earth Day as an opportunity to create measurable changes for the environment. There are now fund-raising walks on Earth Day, such as the March for the Parks, as well as corporate sponsored tree-planting events and numerous roadside and beach cleanups. Because of its national profile, the event is spurring individuals and groups to exploit Earth Day—if you will—by launching various efforts tied to the day. And through all these actions, political pressure can be applied, because as the saying goes, 'When the people lead, the politicians will follow.'"

To further its impact, Earth Day USA hosts an annual conference for all interested parties to increase involvement in Earth Day activities and develop new ideas. The group also publishes a periodic newsletter that helps members share ideas and promotes the various services provided by the organization. As each Earth Day approaches, the number of people involved in the group's efforts grows exponentially. The effort starts ballooning in September, and after January 1 it becomes a fast-paced affair until April 22. During that period, Earth Day USA typically expands from one central office to several regional headquarters around the country.

"For a number of reasons, Earth Day has a high burnout rate," Anderson explains. "For the most part, organizers are volunteers working with limited budgets. There is a high level of intensity around 'the day,' and many times expectations are higher than actual results because to a certain extent, Earth Day activities are still education-oriented and don't always produce measurable results."

Nonetheless, Earth Day USA continues its quest to help the event "take on a life of its own," Anderson is quick to add. "We want people to feel it is a commonly owned event—that it is mine and my neighbor's, and not tied to a specific group." Outside the United States, Earth Day's impact also continues to grow on an international scale. Organizations committed to celebrating the day on an annual basis now exist in Canada, China, Germany, Japan, Russia, and throughout the Caribbean, among other places.

"Our goal is to have Earth Day become an annual occurrence due to the fact that its observance will have become second nature for so many millions of Americans," Anderson says. "By simply existing, Earth Day USA plays a major role in making the day a nationally visible event. Our slogan, 'More than just a day—a way of life,' is a call for every individual and organization to integrate their concern for the environment into their everyday decisions. And to that end, Earth Day itself continues to have an impact on the environment."

Each April 22, inspired people everywhere plant thousands of trees, pick up trash along hundreds of miles of highways, and remind millions of others through the media and various Earth Day education efforts that the environment continues to play a crucial role in the future of humanity. The hope is that by reminding people worldwide about the importance of the environment every April, overall awareness will become more widespread and deeply ingrained and will permeate people's daily actions throughout the rest of the year. Earth Day revolves around people and the power they can generate for environmental protection and restoration.

FOR MORE INFORMATION
Earth Day USA, P.O. Box 470 Peterborough, NH 03458; (603) 924-7720

Children and young adults are becoming increasingly aware of the need to protect the environment. Here, teen entrepreneur Casey Golden proudly displays his biodegradable invention: the Bio-Tee. (Photograph courtesy of Biodynamics, Ltd. Used with permission.)

CHAPTER 7

KIDS FACE THE FUTURE

Unlike many adults, some children have a very different view of the future. A good percentage of adults don't have to worry about the distant future and what shape the environment might be in—they won't be around to live in it. To a certain extent, what the future holds does not concern them. For today's children, however, that isn't true. Some are realizing how important the environment might be to their very existence and are working to ensure that preserved ecosystems prevail.

Teenager Melissa Poe didn't set out to change the world, but her efforts might end up having that effect. In 1989, when she was just nine years old, Poe saw an episode of the late Michael Landon's *Highway to Heaven* television show. The program depicted what the world's environment might have looked like twenty years hence. Pollution was all-pervasive, and ecosystems were in shambles. The images frightened her.

"In twenty years, I figured I'd be just thirty years old," she says. "So I sat down that night and wrote a letter to the president, asking that something be done." When she didn't receive a timely response from then-president George Bush, Poe set in motion a media campaign to make her point that the environment will be important to the future of civilization. That effort ended up bringing her fame and has resulted in a far-reaching environmental effort.

First, she called several outdoor advertising companies and asked about a possible donation of free billboard space. Soon her letter filled a billboard in Washington, D.C., as it also did one in her hometown of Nashville, Tennessee. It read: "Dear Mr. President: Please do something about pollution. I want to keep on living until I'm 100 years old. If you ignore this letter, we'll all die of pollution. Please help."

Twelve weeks later, Poe received a response to her letter, but it was not what she had expected from the president of the United States. The reply was simply a standard form letter asking her to stay in school and not to use drugs.

Poe placed another phone call to the Outdoor Advertising Association of America that spurred the company to feature Poe's letter on 250 billboards across the nation. After appearing on NBC's *Today Show* and talking about President Bush's response, a second letter from the president soon followed, saying he regretted not answering her concerns in the first letter.

When several Nashville television stations also interviewed Poe, other children started calling her. They wanted to know what they could do to help. She decided to start a club called Kids For A Clean Environment (Kids FACE)—to help children get involved in environmental activities and form groups in their schools. Still more children called after seeing the billboards and Poe's *Today Show* appearance. Kids FACE now has more than 200,000 members that undertake tree-planting and recycling projects, among other efforts, and help create more awareness about protecting the environment among their classmates and parents.

After expenses for the club—for such things as postage and long-distance phone calls—grew too large for her family to handle, Poe again relied on her initiative. She wrote a letter to the retail giant Wal-Mart asking for help. Sam Walton, who founded Wal-Mart, wrote back and offered to pay for a bimonthly newsletter for Kids FACE, as well as all related postage. Poe now sends material for each newsletter to Wal-Mart, which designs, prints, and mails it to all the club's members. Material for the publication comes from the more than two hundred letters Poe receives every day, most filled with thoughts, ideas, and poems about the environment.

For her efforts, Poe has become a celebrity of sorts. She was flown out to California to meet Michael Landon and has been featured in a Wal-Mart television ad. During a 1995 Earth Day event in Washington, D.C., she spoke in front of more than 100,000 people. While they were in the nation's capital, Poe and some of her fellow Kids FACE members also unfurled a huge flag the club had created, composed of twenty thousand individual drawings. The founder of Earth Day, Gaylord Nelson, spoke at the flag's unveiling on the Washington Mall, and a letter from Vice President Al Gore was read to the gathering.

The idea for the flag came from a project Poe had initiated earlier for her school's Kids FACE chapter. The club wanted a piece of artwork for its T-shirts, so each member created a drawing on a twelve-inch-square piece of cloth. "We used markers or paints and drew what we thought the world should look like or animals or something involving the earth," Poe says. "Then we picked out our favorite for the T-shirts, but we didn't want the other drawings to go to waste, so we decided to start making a flag."

With twenty drawings composing the Kids FACE flag, the group decided to turn the project into a large-scale effort. Letters from the club about the flag soon spurred articles in periodicals such as *Seventeen, Weekly Reader,* and *Scholastic Magazine.* The Kids FACE newsletter also trumpeted the effort. Over the next year, five thousand cloth squares arrived at the club's headquarters in Nashville, each carefully created by children around the world, Poe said. The squares featured sayings, such as "Help save the earth," or poems or song lyrics or pictures of eagles, trees, or other environmental images.

When a final call for squares went out in the club's newsletter early in 1995—along with information about dedicating the flag on Earth Day in Washington—yard after yard of material soon arrived at Poe's doorstep. More than fifteen thousand squares tripled the size of the flag in just over three months, and Poe, her mother, and other volunteers had plenty of sewing to do in order to meet their deadline. "I was amazed at how big it became," Poe said. "It was exciting to watch it grow bigger and bigger."

By Earth Day, the flag measured one hundred feet by two hundred feet. More cloth squares continued to arrive after the dedication ceremony, and Poe says her club's flag—along with its overall mission to involve children in environmental activities—will continue to grow, too. "Kids FACE is there to get kids involved in helping the environment," Poe says. "I hope people start to realize that kids can make a difference when they put their minds to it. One day, hopefully the environment will be OK, and we won't have to worry about it anymore."

The same concern for the future of the environment that drove Poe to start her club is spurring other children to initiate entrepreneurial ventures. These projects aim at making a positive impact on the environment while also forming viable businesses. In one such venture, a sixteen-year-old entrepreneur from Evergreen, Colorado, has developed a biodegradable golf tee made of recycled materials in an effort to make one aspect of the popular sport environmentally friendly.

Although golf tees are small, their numbers add up: Each year, more than 1.5 *billion* golf tees are used in the sport—the equivalent of thirty-five thousand mature birch trees. "I knew if I used recycled materials, the tees would help save trees and stop those materials from going to landfills," says Casey Golden, creator of the Bio-Tee.

The idea for this new form of golf tee came indirectly from Casey's father, who used to work as a greens-keeper and knows what broken tees can do to lawn mowers. "Whenever I went golfing with my dad, he made me pick up the broken tees on the tee boxes," Casey said. "Finally I said, 'Why can't these things just go away naturally.'"

When a project at school asked students to think of a problem and come up with a creative solution, Casey—then thirteen years old—went to work creating a biodegradable tee. His original formula included flour, water, fertilizer, peat moss, and applesauce. The unusual mixture was then pressed into the tips of caulk containers and hardened in the microwave oven. Casey entered his idea in the Invent America contest, a national competition sponsored by the U.S. Commerce Department for kindergarten through eighth grade students. He won the thousand-dollar grand prize for his grade level.

Now, patented formula no. 56 for the Bio-Tee is composed of recycled paper, recycled plant fibers, water-soluble binders, and a pulplike by-product of the beverage industry. Golfers won't notice a difference when using a Bio-Tee to tee up, but once one is used, it soaks up water when the sprinklers come on. The next mowing then disperses the material as mulch. If Bio-Tees were used exclusively in the United States, nearly a million pounds of waste would be diverted from landfills every year.

Casey's father, John Golden, is the president of Bio-Dynamics, Ltd., a company the family started to produce and market the Bio-Tee. Numerous thirty-two-cavity molds produce batches of the tees every thirty-five seconds. The round-the-clock operation produces seventy-five thousand tees a day, and orders continue to roll in. Other manufacturing alternatives are being explored.

Bio-Tees are now being sold in Kmart stores across the country. The company has featured Casey on the cover of one of its quarterly reports, honoring both his entrepreneurial skills and environmental awareness. "It's been interesting," Casey says of the path his idea has taken. "You'd think it would be easy to just make golf tees, but it takes a lot more effort than I first thought."

Along with their actions and enthusiasm for the environment, children's education can play a big role in their concern for the planet. To Wyoming entrepreneur and ex-teacher Thayer Shafer, helping children understand the earth and all its intricate systems is one of the most important things that can be done to help the environment. To that end, Shafer formed GeoLearning, a company offering hands-on educational materials and science projects that combine fun and student involvement with learning about the earth and its environment.

Although the materials can be incorporated into games, their primary purpose is to educate children about the planet's unique natural systems, such as its oceans and continents. American children have performed poorly in geography compared to students from other countries. Grasping the proper perspective of the earth—its land masses and water bodies—is an important first step in understanding global environmental issues. For instance, one unusual GeoLearning map shows the oceans as the single water body they in actuality are. By seeing that the world's oceans are interconnected, children realize that pollution in one body of water can affect the others. A better understanding of the physical makeup of the earth will lead children to have a greater concern for the environment.

"The biggest single thing is education—to teach kids so they understand the earth and realize we're on a unique and dynamic planet that is continually changing," Shafer says. "Understanding the planet helps people understand that you don't mess with Mother Earth and that you have to treat the planet with care."

Many of GeoLearning's products were created by Athelson Spilhaus over the past fifty years. Spilhaus, an oceanographer and a science advisor to presidents Eisenhower, Kennedy, Johnson, and Ford, invented new map projections—new ways to look at the earth—and adapted them for children. Shafer purchased the oceanographer's prototypes, refined them, and eventually had them produced to sell to teachers around the country when he started GeoLearning.

The company's most popular product is the Geoglyph, Latin for "earth carving." It is a puzzle composed of nine pieces, each one either a continent or an ocean. The pieces can fit together in more than one hundred correct combinations. The puzzle is a great learning tool for children because they are always right, Shafer says. Psychologists have even used the product as a therapy tool for patients with a fear of failure. Other products include a 470-piece magnetic globe puzzle that features the world's countries as well as a movable "hexaflexagon" that demonstrates the concepts of sea-floor spreading and continental drift as children spin through the different phases.

GeoLearning aims its marketing at schools, home schoolers, parents, and children. It also sells its products at science and children's museums as well as in map stores. The Epcot Center in Orlando, Florida, distributes hundreds of GeoLearning catalogs every month at a sales site set up strictly for teachers.

"The more children understand the planet," Shafer says, "the more they will work with it as they grow up instead of fighting against it as we have done in the past." In other words, by grasping how the planet's various ecosystems rely on one another and work together to create a hospitable environment for all of life, children will understand the need to protect those ecosystems instead of following the tradition of over-using and polluting our natural resources, in many cases without concern for the future. With decades of life ahead of them, the future is everything to children.

FOR MORE INFORMATION

Bio-Dynamics, Ltd., P.O. Box 2013, Evergreen, CO 80437-2013

GeoLearning, P.O. Box 711, Sheridan, WY 82801; (307) 674-6436

Kids FACE, P.O. Box 158254, Nashville, TN 37215; (800) 952-3223

CHAPTER 8

HELPING CULTURES SURVIVE

Trees aren't the only things being exterminated in the world's forests. Ways of life, too, are being lost.

Repeatedly over the years, native peoples have lost control of their lives and land in the name of development and progress, which in turn has led to many abuses of the ecosystems they once cherished and protected. Worldwide, more than 150 million people have been displaced from lands they traditionally occupied. For instance, in Sarawak, the Malay-owned portion of the island of Borneo, loggers have cut down a third of the rain forest. Only 450 native Penan people remain where ten thousand lived before logging began.

These human rights abuses and related environmental disasters date back centuries to the colonial period, when European nations traveled the globe and gained control of countries and territories in Asia, Africa, Latin America, and the South Pacific. In most cases, once control was wrested from the native peoples of these lands, they became second-class citizens at best and were invariably viewed as expendable by their European conquerors. In the name of development and progress, the new rulers set up governments that justified the exploitation, displacement, and even elimination of native peoples who dared question their motives and actions, which almost always revolved around material gain.

Sadly, modern examples of these colonial practices still exist in numerous Third World societies today. For instance, the United States has joined the fray with its own special brand of colonialism, one that relies on multinational corporations that take advantage of native peoples. For example, U.S.-based oil and logging companies operate in jungles and rain forests around the world with little or no concern for the environment or local peoples. Because vestiges of the colonial past remain in many developing nations in the form of their opportunistic government officials, these operations flourish. Indigenous peoples are still contending with colonial rule because these large corporations, working with these government officials, wield huge amounts of power over the lands of native peoples.

However, in some cases today these quasi-colonial practices are being stopped. This is due in part to the efforts of an organization called Cultural Survival. Helping indigenous people and ethnic minorities deal as equals in their encounters with industrial society is the task this nonprofit group laid out for itself in 1972. It has since supported the efforts of endangered cultures around the world as these diverse peoples fight to save their traditional ways of life. Although Cultural Survival's efforts revolve around human rights, the environment plays a significant role in the group's work as well.

"There is a definite connection between the two," asserts Michelle McKinley, director of Cultural Survival. "Environmental injustices happen to indigenous communities that don't have clout—people whose human rights simply are ignored. For example, petroleum companies working in the Amazon River Basin in South America are allowed to use old equipment in poor working condition that dumps oil and fouls the environment. The companies get away with it because the indigenous people can't fight back. They are not recognized by their own governments. Securing people's rights to be there is a way of taking care of the environment."

In too many cases today, indigenous societies continue to be rendered powerless by many of the same forces that came into existence in colonial times. Because many of these cultures are rooted in remote areas and do not participate in their countries' modern society, native peoples are all too often considered expendable citizens. In being forced to deal with the modern world to secure their rights, indigenous societies are ill prepared. Their "poverty" and "illiteracy," value judgments applied to their traditional ways of life by the modern world, hamper their ability to fight for their rights. By helping secure indigenous societies' rights to maintain control of their native lands, Cultural Survival is working to protect both human rights and the environment. With control of their lands in their own hands, native peoples will work to preserve the ecosystems that have sustained them for centuries, and if there is development, it will be on *their* terms.

Cultural Survival's birth grew out of the initial development efforts in the Amazonian rain forests, which became possible thanks to terms set up by the Brazilian government. In the late 1960s, the government of Brazil began opening the country's vast tracts of land to national and international business interests. At the time, anthropologist and Cultural Survival's founder David Maybury-Lewis was in Brazil training scientists to work with and study the indigenous people of the country's rain forests. He foresaw the misery and helplessness that lay ahead for many of the native people and created Cultural Survival by combining academics with advocacy to try and save the world's numerous native cultures. Initially, the organization applied for grants to conduct small-scale health and development projects, but within a decade, indigenous people and communities began organizing themselves on their own. The nonprofit group then focused on supporting these groups' efforts and fostering their autonomy.

Today, Cultural Survival assists these communities by giving them technical assistance in learning how to secure land rights, build effective organizations, and manage their natural resources. More than fifty projects are under way on five continents. "A key element is making sure these peoples' land bases are secure, so we spend a lot of time working on land-title issues," McKinley says. "But these groups must decide for themselves what traditions to maintain—and how fast to change others—as they encounter majority cultures and state governments. Along with working with the native groups, our effort involves educating the public, policymakers, and governments about indigenous issues. Governments need to make sure their review processes for development projects include the indigenous communities that will be affected."

Cultural Survival's success stories are growing. For example, in 1988, after gold was discovered in the native lands of Brazil's Yanomami Indians, more than fifty thousand miners flooded the area. They brought diseases the Indians couldn't ward off, poisoned their water with mercury used in gold mining, and killed those Indians who stood in their way. By supplying the Yanomami with medical and organizational support, Cultural Survival aided the Indians in their battle to expel the gold miners. In 1991, the group helped the Yanomami receive full title to their land from the Brazilian government. Although the land was obviously valuable in terms of minerals, in this case the government decided to do the right thing and grant the land to its rightful owners—albeit at the urging of numerous human rights and environmental organizations.

In other efforts, Cultural Survival is helping the Awa Indians in parts of Ecuador and Colombia map and gain title to their lands while managing their natural resources in the world's wettest rain forest. Cultural Survival also supports a number of native farming and foraging cultures in Africa and Asia. The organization's *States of the People*, a 260-page document originally published in 1993, lists more than 250 of the world's indigenous and ethnic societies. It provides census, societal, and distribution information about the groups along with news about land being contested and other struggles. Cultural Survival plans to make *States of the People* available over the Internet to provide wider access to the information it contains. Interested parties and individuals wanting to help the societies listed will then have easy access to information that can be useful in fund-raising or other advocacy efforts.

In another far-reaching endeavor, Conservation International is helping the environment and native peoples by linking conservation efforts to local economic needs. In building markets for products harvested by communities in sustainable ways, the nonprofit U.S. organization is creating defenders of rain forests and other threatened ecosystems. Thanks to one of the group's ongoing programs, tagua nuts from South American rain forests are being used to make ivorylike buttons for more than thirty clothing manufacturers around the world. Called the Tagua Initiative, the effort requires a complicated mix of diverse fields, including biology, business administration, community development, and conservation planning, says Robin Frank, director of the Ecuador-based project. "Deforestation of the rain forests is partially driven by a lack of employment alternatives for the local people. We want to offer viable economic alternatives for the long run that will help the local people and save these ecosystems."

More than 30 million tagua buttons have already been sold. The more than 1,500 tons of tagua used has generated more than $3 million in button sales, and the local Ecuadorans are receiving their fair share. Tagua was a popular button material in the early years of this century until inexpensive plastic took its place. At one point in the 1930s, one in five buttons manufactured in the United States was made from tagua. The Tagua Initiative first attracted sportswear makers Patagonia and Smith & Hawken Ltd., who wanted the alternative buttons for their clothing lines. Other companies, such as Esprit, The Gap, Banana Republic, and Timberland, have since boosted the effort with more tagua-button purchases.

Tagua palm trees grow throughout western South America, but the species with the highest quality nuts grows only in northwestern Ecuador. The nuts are harvested by local people, then dried and sliced before being shipped to button manufacturers for final processing. Tagua Initiative employs more than eighteen-hundred Ecuadorans in jobs ranging from collecting the nuts to handcrafting buttons. Because the golf ball–size nuts resemble elephant ivory in texture and appearance, markets for tagua jewelry, chess pieces, and carvings are opening up. An artistry training center has been set up in Ecuador to teach local artisans how to create tagua carvings of endangered animal species and other subjects. Men who once cut down tagua palms are now protecting the trees so they can sell nuts. Many local Ecuadorans have quit working as loggers because they can make better livings by joining the initiative.

To ensure the Tagua Initiative and other ventures such as these accomplish their goals, community development must be tied to scientific research to protect the rain forest from possible damage, Frank explains. This means relying on biology and other disciplines to make sure the environment is safeguarded. For instance, if too many nuts or seeds are harvested, ecosystems could be damaged. Or if various gathering and production methods lead to other forms of ecological damage, the projects' ultimate goals of gaining economic benefits while protecting the environment are defeated.

"The whole management issue is crucial," Frank asserts. "We must harvest the rain forest using a scientific basis or there still will be potential for harm. To that end, ecological monitoring assures that products are collected without damage to the surrounding forest, while social and economic monitoring provides insight into just how great our impact is. In addition, by teaching business management skills at the community level, we're building knowledge so the people can manage these ventures in the long run."

As a working example of alternative land use, the Tagua Initiative has become a focal point of local debates in Ecuador concerning how to manage the forest. Local people rejected a proposed thirty-thousand-acre banana plantation that would have destroyed native rain forest, including tagua trees, because of the initiative's success. "The Tagua Initiative is providing a perfect example of how [this] type of environmental project can work," Frank says. "If we can manage the rain forest and provide income and jobs for the local people year after year, these projects are going to make a difference." Not only do they provide employment that can replace activities such as logging and slash-and-burn farming, which harm the rain forest, but the projects also end up teaching local communities the value of the intact forests in terms of sustainable uses and economic gains.

Conservation International currently has a number of other projects under way in other parts of the Third World. Hundreds of other sustainable rain forest products with economic potential are being researched by the organization and incorporated into new efforts. For instance, one project employs a number of people living in and around a huge rain forest nature reserve in northeastern Gùatemala, near the Yucatan Peninsula. Called the Maya Biosphere Reserve, its 3.7 million acres compose the largest tract of rain forest left in Central America. By forming a partnership with Croda Inc., a for-profit supplier of raw materials for cosmetics, Conservation International is identifying and marketing renewable rain forest materials that can be used in everyday products like soaps, creams, shampoos, and other personal care products. So far, the joint venture is creating financial benefits for local Guatemalan organizations supported by Conservation International. Native employees sustainably harvest and process four raw ingredients for Croda products—efforts that again provide economic alternatives to deforestation, says Sharon Flynn, director of Conservation International's enterprise department.

One of the ingredients used in Croda's products is cohune oil, which is extracted from the nut of the cohune palm. The tree grows abundantly throughout this section of Guatemala, known as the Peten. Local residents harvest the nuts, remove the seeds, and press them to release the oil, which is then shipped to Croda for refinement. The resulting oil can be used as a replacement for coconut or mineral oil, ingredients commonly used in skin care and other personal care products. Another type of oil is derived from the allspice tree. Leaves from the trees are dried and sent to Croda, where they undergo a rinsing process to remove the fragrant oil, which can be used in lip balms and skin creams.

A third material from Guatemala is jaboncilla extract, obtained from the dried red berries of the jaboncilla plant. The rose-colored liquid extract foams in water and was used as a soap by the ancient Maya. Today the extract can be used in the cosmetics industry for facial washes and other types of soaps. A fourth material involved in the joint venture comes from Peru. Brazil nuts are harvested by local people in the Peruvian Amazon, then shelled and pressed. The crude Brazil nut oil is then shipped to the United States, where Croda chemists refine it and conduct animal-free safety tests on it. Croda eventually sells the Brazil nut oil, along with the three Guatemalan materials, to cosmetics manufacturers.

Yet another Conservation International effort in Guatemala involves gathering dead leaves and other rain forest debris for a new line of potpourri products called Gatherings, which is sold in a number of U.S. department and specialty stores. A community of twelve hundred in Guatemala benefits from this enterprise.

"Collecting nuts and berries won't save the rain forest in and of itself, but the efforts are tools for conservation," Flynn says. "By creating jobs and demonstrating how the rain forest can be used sustainably and left intact, we are supporting other strategies that can make a difference, like education, land-rights issues, and other community conservation efforts."

New logging practices aimed at replacing large-scale industrial efforts in Guatemala are also being supported by Conservation International. Recognizing that logging will continue in areas where it is profitable, the organization is supporting ecologically sound methods of harvesting wood in locations already committed to timber extraction, explains Jim Nations, vice president for Mexico and Central America programs. Studies have been completed showing that these small-scale logging efforts can work on a sustainable basis, and plans have been drafted to start up several of these ventures. However, the Guatemalan government has yet to make a decision on whether it will allow the existence of these smaller logging operations.

"In the past, the government gave large concessions to big companies that came in, built roads, took out all the high-grade trees, and left," Nations says. "An alternative is to give smaller concessions to people living in the forests and let them work on community-based sustainable forestry efforts. Because they have a spiritual connection to the land and know their water and other products come from the forest, they will keep it alive." A number of these smaller logging efforts could harvest the same amount of lumber as a large industrial concession. But with the smaller operations, both the forest and its inhabitants would be protected and preserved.

Of course, although all these various sustainable efforts benefit the environment and the native peoples participating in them, they are still limited in the cumulative positive impact they can provide for the earth's indigenous cultures and natural world. In too many cases, native peoples still are left powerless in their struggles to protect their homelands or have already fled the land to try and survive in large cities. Nonetheless, these ventures provide hope that more sustainable alternatives can be created that will preserve the integrity of native peoples, allow them to maintain their spiritual connection to their lands, and protect the environment.

FOR MORE INFORMATION

Conservation International, 1015 18th Street NW, Suite 1000, Washington, DC 20036; (202) 429-5660

Cultural Survival, 96 Mt. Auburn, Cambridge, MA 02138; (617) 441-5400

Politics often plays a role in protecting or compromising the environment. The Southern Utah Wilderness Alliance is one of thirty-six organizations working to ensure the preservation of Utah's natural wonders from government-sanctioned development. Here, hikers work their way down Forty Mile Canyon on U.S. Bureau of Land Management land in the Escalante River drainage.
(Photograph by Kevin Graham.)

Building Political Power for the Environment

Most, if not all, politicians claim they want to protect the environment. Fortunately, some organizations don't take their word for it but work to protect the environment *from* politicians.

Politics plays a huge and crucial role in protecting or sacrificing the environment. The U.S. Congress tackled serious environmental problems in the early 1970s after a public outcry arose over a polluted river that caught fire, thick air pollution that hung over many metropolitan areas, and trash that covered landscapes across the country. This outrage stirred Congress to quickly enact legislation to address these concerns, and the political body even went on to create the Environmental Protection Agency with its mandate to continue protecting the environment.

These political actions involving the environment, along with numerous subsequent efforts, have played a critical role in protecting the natural world and human health. For example, compare U.S. legislation concerning the nuclear power industry to the former Soviet Union. Although the Three Mile Island nuclear accident in Pennsylvania was not a high point in U.S. history, it pales in comparison to the Chernobyl nuclear disaster in the Ukraine, which will be wreaking environmental and health catastrophes for years to come. Strict regulation involving a potentially dangerous source of power, as is found in the United States, could have prevented the Chernobyl nightmare.

Another example can be found in the hundreds of toxic Superfund sites scattered across America. These are the results of business and military practices not held in check by law. It is no secret that businesses of all kinds will take advantage of the environment in order to turn higher profits. This is the dark side of capitalism.

Many people argue that the government has too much control when it comes to telling businesses and individuals what they can and cannot do in terms of the environment. Needless regulation can sap profits and cost jobs, they argue. The key is to try and find a balance between protecting the environment and allowing for the effective operation of a capitalistic system. Many people on both sides of the issue would argue that finding such a balance is impossible. But the fact remains that reaching a balance is the only potential solution we have available at this time, and it is the very reason we elect politicians to represent us in Washington, D.C.

There are, of course, plenty of environmental groups working in the political world to help the earth in any number of ways. However, at the present time only one works exclusively at the task of getting proenvironment candidates elected to the U.S. Congress—the League of Conservation Voters. The league's mission is simple: Get a bipartisan majority of candidates who support the environment elected to the U.S. Senate and House of Representatives. "The idea is to be the political arm of the U.S. environmental movement. That's what we've been since 1970, when leaders of the environmental movement decided they needed to try and impact Congress," says Sarah Anderson, the league's research director. "We use the tools of political campaigners to get politicians into Congress who will vote for the environment. And once we help get candidates elected, we hold them accountable for their conservation voting records."

To that end, each year the league produces a National Environmental Scorecard that rates all members of Congress on their support of environmental protection. The Scorecard is sent to the group's more than thirty thousand members—U.S. citizens of all creeds and races—as well as the media. It includes factual information about the environmental votes that faced the past Congress and details the environmental voting records, or scores, for each legislator, state delegation, and region of the country. The congressional votes, which are tied to specific bills, are tallied on each Scorecard after being determined by experts from twenty-seven large U.S. environmental organizations. These experts meet and decide which votes in Congress are most crucial to the nation's environment. For example, the Environmental Scorecard summing up the work of the 103rd Congress in 1994 considered votes on such issues as food safety, safe drinking water, renewable energy, toxic military bases, environmentally safe technologies, national parks, and wilderness.

A score ranging from 0 to 100 percent—with 100 being a perfect score—is then registered for each member of Congress based on his or her voting record on the list of bills chosen for the Scorecard. Overall, Democrats in the 103rd Congress averaged scores of better than 70 percent while Republicans managed overall average scores of 19 percent in both the House and Senate—not surprising, because

Republicans tend to be prodevelopment and shirk from regulation. The New England and Middle Atlantic regions scored the highest, and the Rocky Mountain and Southwest regions tallied the lowest scores. Politicians from the Rocky Mountain region scored lowest overall: 28 percent in the Senate and 31 percent in the House. This part of the country is notorious for electing prodevelopment politicians that back the aims of the mining, logging, and cattle industries and tend to think of the wide-open lands of the West as ripe for profit making.

In its efforts to get proenvironmental politicians elected, the league uses the tools of all successful political campaigns. It helps run phone banks, coordinates mailings, and trains campaign volunteers. The organization even created a program called EarthList to collect cash donations for candidates it supports. Individual members of the league can choose a candidate they wish to support with a cash donation, write a check to that person, and send it to the league. The league then collects all the checks and delivers them to the candidate to show that environmentalists are together in support of his or her candidacy. "It gives clout to the individual who writes a small check, because the donation is put together with a lot of other individual checks," Anderson says. "No matter what environmental organization you belong to, it's good to remember that those groups don't do politics, per se. They lobby on Capitol Hill to the politicians who end up there, but we're actually trying to change the collection of politicians who represent us in Washington to reflect the proenvironmental concerns of the American public."

The league estimates that more than $50 million a year is spent to push for a proenvironmental agenda in Congress, if one considers all the money spent by the vast array of environmental groups and their members. But just a fraction of 1 percent of that amount is spent on trying to get proenvironmental candidates elected to office. In addition, the league estimates it is being outspent by prodevelopment special interest groups by at least a thousand to one in election spending.

"Right now, too often politicians are voting for the short-term profits of the industries that fund their campaigns—and that's an easy vote for them to cast," she said. "Then industry can point out how certain legislation will cost a lot of money and ask that the politician put off a vote on it for a few years. With the funding that has been provided, that's an easy call for a politician. We're a force in the political arena that can attempt to match these other forces for short-term profits at the expense of the environment."

Far from Capitol Hill, a Utah nonprofit group is concentrating on political efforts aimed at preserving an ecosystem dear to the hearts of its more than twelve thousand members. Through a combination of research, public education, grassroots activism, legislation, and litigation, the Southern Utah Wilderness Alliance (SUWA) is dedicated to making sure the desert wildlands of this western state are protected. Formed in the early 1980s, the group's main goal revolves around a congressional bill that would protect 5.7 million acres of southern Utah public land as part of the National Wilderness Preservation System, says SUWA executive director Mike Matz. The public lands surrounding Utah's spectacular national parks are characterized by equally impressive arches, buttes, canyons, pinnacles, and large expanses of slickrock—all composed of varying shades of brilliant orange and red sandstone.

The U.S. Bureau of Land Management (BLM), however, has proposed that only 1.9 million acres be considered for wilderness designation out of the 22 million acres of BLM land in the state. To fight for more wilderness, SUWA has joined with thirty-five other organizations to form the Utah Wilderness Coalition. The wilderness proposal pursued by the coalition would protect numerous archaeological sites, pristine canyon systems, critical habitat, and migratory routes for wildlife from development.

The proposal has been sponsored in the U.S. House of Representatives by an environmental congressman, Representative Maurice Hinchey of New York, and has been cosponsored by numerous other representatives. However, a competing proposal introduced by Utah's congressional delegation in 1995 would designate only nominal acreage as wilderness, Matz says, and would preclude the remaining land from ever being designated wilderness by Congress, thus opening up huge tracts of the land to development.

"We're in the midst of a public-education campaign to broaden support nationally for the broader wilderness proposal. Because these are federal public lands, we are trying to reach out to people across the country who know the wonders of Utah's Colorado Plateau and create a national effort. Unless we succeed, what remains of Utah's spectacular wild lands will be handed over to extractive industries, such as mining. America's red-rock wilderness deserves better."

SUWA funds slide presentations about the proposed wilderness lands around the country. A former executive director of the group heads the effort. A dozen SUWA staff members operate four offices: one in Washington, D.C., for the group's lobbying efforts, and three more in Utah—Salt Lake City, Moab, and Cedar City. A Salt Lake City newspaper once called the group the state's most influential environmental organization, in large part because of SUWA's talented staff and active membership. "SUWA protects public lands in Utah for the types of values that are often ignored," Matz says. "We are advocates for recreation, wildlife, and healthy ecosystems."

Sadly, even recreation is becoming a development issue in this part of Utah as sections of the desert country run the risk of being loved to death. For example, the town of Moab is now a mecca for both four-wheel-drive enthusiasts and mountain bikers. Both activities, if unchecked, can wreak havoc on desert landscapes. Tires of all types destroy fragile desert-soil ecosystems, which can lead to extreme erosion problems. Wilderness designation, however, will provide regulation and monitoring to further protect the land.

All SUWA members receive a quarterly magazine updating them on progress and setbacks on many fronts of the struggle to preserve Utah's canyon country. The publication also gives members opportunities to get involved by encouraging letter-writing and phone-call campaigns aimed at various politicians. A section called "Write Time" encourages members to "choose at least one of the following issues to add your voice to the protection of canyon country." Numerous short articles update readers about news on the area's national forests, well-known canyons, government agencies, and industry development plans.

Profiles of SUWA's board of directors also are featured. New Mexico's Jim Baca is a recent addition to the group's board. Baca served as the director of the Bureau of Land Management in the Clinton administration until he ran into political trouble because of his push to reform public-land policies, including grazing and mining programs. His reform ideas, which would have curbed the subsidies now provided through these programs, were seen as too radical.

"As far as wilderness is concerned, there is no more important environmental fight than the fight for southern Utah," Baca said in the SUWA publication. "I believe in organizations like SUWA and their ability to garner real grassroots support. Without grassroots involvement, we cannot expect to win the battles, and southern Utah is going to be wasted." Baca feels strongly about this wilderness battle because of the desert lands in question. They offer a stunning form of beauty not found in any other part of the country and also contain hundreds—if not thousands—of ancient Indian dwellings and ceremonial sites. Once opened up for development, much of this land will face the risk of being altered forever.

On the other side of the country, a nonprofit organization based in New York City that initially grew out of a singular environmental tragedy is dealing with political issues on a global scale. Upset by the 1987 Chernobyl nuclear disaster in what was then the Soviet Union, a small group of professionals gathered to determine what, if anything, they could do to help slow the destruction of the world's environment. Led by Dr. Christine Durbak, a psychoanalyst from New Jersey, the group eventually formed the World Information Transfer (WIT), a nongovernmental organization affiliated with the United Nations. Along with its New York headquarters, the effort also has regional offices in six other cities around the world.

"We wanted to create a way to educate political leaders, opinion makers, and citizens of the long-term consequences of environmental disasters and degradation," says Dr. Claudia Strauss, a member of WIT's board of directors. "We recognized the critical need to provide information about our deteriorating global environment." Strauss, who has taught at the university level, says WIT is simply a way of educating people on a different level. By collecting and disseminating environmental information, the group hopes to increase awareness, stimulate debate, and spur corrective actions. To accomplish this, WIT sponsors seminars, broadcasts, and lectures and produces a quarterly publication, *World Ecology Report*, a digest of current international environmental health news. Each issue tackles a specific area of environmental concern, such as clean water, population control, or sustainable development. A Chernobyl update also is included in every issue of the *Report*, which is published in six languages.

Additionally, the group sponsors an annual international conference, Health and Environment: Global Partners for Global Solutions, at the United Nations headquarters in New York City. These public events bring together world leaders and medical authorities to discuss environmental issues and promote responsible solutions to health problems related to continuing environmental contamination. WIT recently created a thirty-minute video featuring the major presentations given at each conference. It has been broadcast on television stations in New York City, Connecticut, and New Jersey, as well as on radio in Kiev, the capital of Ukraine, which is near the site of the Chernobyl nuclear power station.

To further its information-gathering efforts, WIT has opened a research institute in Kiev. Called the Center for Creative Sustainability Studies, the facility focuses on collecting information on environmental issues in the region—specifically, nuclear issues involving the Chernobyl disaster and its aftermath. "We see ourselves as an environmental education organization hoping to change industry's and people's behavior toward the environment," Strauss says. "We want to influence politicians, educators, and the media—to get them to recognize the impact on human health caused by environmental degradation. Our underlying premise is that knowledge about health and the environment is the surest way to prevent other man-made disasters that are so devastating to human health."

Although following politics and all its wrangling can be trying and at times downright depressing, legislative actions continue to play a huge role in the future of the environment. And although you may feel helpless in trying to deal directly with problems like the disappearing ozone layer or rain forests, you can make an impact on these issues and many more through politics. Many environmental groups, such as the Sierra Club and Wilderness Society, use part of their members' dues to conduct lobbying efforts in Washington, D.C., on behalf of environmental issues. By joining any of these groups you are taking political action.

Another important political measure involves nothing more than a phone call or two. Short calls to your congressional representative and senators can potentially sway their votes on certain environmental issues—particularly if other constituents are voicing similar concerns. A bolder step is writing a letter to the editor of your local paper. Politicians pay attention to these letters as a way of checking the public's pulse on various issues. And when you enter the voting booth every November, you make a political statement for or against the environment with nearly every lever you pull or hole you punch.

FOR MORE INFORMATION

League of Conservation Voters, 1707 L Street NW, Suite 750, Washington, DC 20036; (202) 785-8683

Southern Utah Wilderness Alliance, 1471 South 1100 East, Salt Lake City, UT 84105-2423; (801) 486-3161

World Information Transfer, 444 Park Avenue South, Suite 1202, New York, NY 10016; (212) 686-2929

Conservation tourism, or "ecotourism," means visiting a place, learning about it, and then leaving it intact. Here, Earthwatch ecotourist volunteers help to build a solar oven in Kenya. (Photograph courtesy of Kammen/Earthwatch. Used with permission.)

CHAPTER 10

UNITING CONSERVATION
AND TOURISM

There is a fine line between ecotourism and ecodestruction. A delicate balancing act exists in preserving the earth's most beautiful places while also introducing these natural treasures to curious travelers. As more and more companies join the burgeoning field of ecotourism, exploitation of both natural resources and indigenous peoples becomes a greater concern, for some of these firms claim to embrace the tenets of ecotourism but don't. In its purest form, ecotourism offers responsible travel and sightseeing activities that conserve the environment by sustaining the well-being of the local people and not harming the precious places and cultures involved. It means visiting a place, learning about it, and leaving it intact.

Conducted properly, this form of tourism can play a critical role in conservation and preservation efforts by promoting public awareness and understanding. However, as the ecotourism business has exploded in popularity, some tour companies have appropriated the name to drum up business while ignoring the conservation aims of this new form of tourism.

The nonprofit group that set the tone for this new industry existed long before the term "ecotourism" became common. Earthwatch has provided meaningful ecological experiences to thousands of travelers since 1972. The organization, which operates something like an environmental Peace Corps, sends out teams of volunteers to invest time and energy in various environmental efforts while enjoying the uniqueness of different cultures and ecosystems.

Projects range from saving endangered black rhinos in Zimbabwe to creating management plans for Brazil's endangered rain forests to studying the effects of tourism on a U.S. national park. Specific Earthwatch successes include saving thousands of endangered sea turtle hatchlings on the Caribbean island of St. Croix, discovering new species of rain forest canopy insects in Peru, and documenting Native American rock art in Utah.

Over the years, the environmental organization has mobilized more than forty thousand people to work for two weeks at a time assisting leading scientists and scholars in researching critical environmental issues. In 1995, Earthwatch sent more than four thousand volunteers into the field to help with 150 projects across the United States and in fifty countries, says Earthwatch President Brian Rosborough. "Earthwatch is about working toward solutions to global concerns," he says. "It's a way of giving people ownership over issues that would otherwise be left to governments and authorities—over which they have little control. We find capital and manpower to go out and address as many of these issues as we can."

One project under way in Kenya has Earthwatch volunteers working with local communities to build solar ovens. Millions of people in developing countries such as Kenya rely on wood and other traditional fuels for more than 50 percent of their energy. Cutting and burning wood contributes to a number of environmental and health problems, including deforestation, loss of biodiversity, and respiratory illnesses. While solar power and other renewable energy sources seem like obvious solutions, they have been used only rarely and have not been sufficiently tested, according to Rosborough. Kenya provides a perfect case study. In three years of Earthwatch-sponsored work, the group's volunteers have helped Kenyan villagers build more than two hundred solar ovens to replace open-fire cooking. Ongoing efforts include construction of windmills and the development of a locally run training center for renewable energy research.

All the organization's projects are listed in each issue of *Earthwatch* magazine, which is sent to the organization's sixty thousand-plus members six times a year. Earthwatch volunteers make a tax-deductible contribution to help support the research effort they participate in. Contributions range from $700 to $2,500 and cover food, accommodations, field support, and equipment. Airfare is not included. Participants, who need no special skills, are trained on-site in teams of five to ten people.

"It's a wonderful way to accomplish a number of different objectives. Apart from assisting scientists who need the volunteer talent to help them do their work, we also do a lot of training of teachers and young people from the host countries," Rosborough says. "By assigning native people to different expeditions, they can bring back knowledge and understanding to share with their own communities in the countries where we are working."

Additionally, Earthwatch is a leader in sponsoring experiential education opportunities for students and educators. Over the years, more than six thousand high school students, along with numerous elementary and secondary school teachers, have received career training as participants in Earthwatch's scholarship and fellowship program. With the help of private donors, foundations, and corporations, the group provided education awards to nearly four hundred students and teachers in 1995, enabling them to join various Earthwatch projects around the world.

Earthwatch has grown along with the entire tourism industry, which is now the largest industry on the planet. For instance, the number of tourists visiting Costa Rica's national parks has grown by 500 percent, from 50,000 to 250,000 travelers, in the last five years. The Ecotourism Society was formed in the early 1990s to monitor this booming industry. It works to help rein in rogue tour operators, educate the industry and tourists on proper conduct and practices, and develop models of sustainable tourism. This Vermont-based nonprofit society promotes tourism as a tool for both conservation and sustainable development in parts of the world where fragile environments are threatened. It also serves as a watchdog effort to help interested tourists determine which tour operators are legitimately involved in ecotourism efforts. Ultimately, it hopes to reduce travel's overall adverse effects on the environment.

Sadly, a number of tour operators around the world don't fully understand all of what ecotourism should encompass, says Megan Epler Wood, the society's executive director. A number of problems involve the mishandling of animals. For example, one tour guide in South America's Amazon region thinks tourists want to see him wrestle the native alligators. Most ecotourists want to see the alligators in their natural environment. They are paying the guide to show them the area's flora, fauna, and people, not to put on a Tarzan show.

In other Third World communities visited by ecotour groups, the guides and native people either capture or attract wild animals for the tourists to pet, feed, and photograph. This practice lowers the animals' defenses if they encounter human hunters later in their lives. It also leads the animals to rely on humans for food, jeopardizing their ability to fend for themselves. Some tour guides set up tours to isolated native communities that are ill-prepared for an onslaught of Western tourists. In Nepal, for instance, native people continue to cut down their dwindling forests in order to boil water for the hordes of tourists who hike in the Himalayas each year.

"Our goal is to further define the term 'ecotourism' so it results in the protection of natural resources and the well-being of local people," Wood says. "More developing nations are seeking to take part in ecotourism activities now because they want to gain the economic benefits and play a major role in protecting their natural resources while promoting tourism. Ecotourism programs must generate conservation benefits for the nations being visited and development alternatives for local people. We are dedicated to finding the resources and building the expertise to make tourism a viable tool for conservation and sustainable development."

The Ecotourism Society acts as a provider of information and training opportunities for tour operators as well as a research facility to study the impacts, both positive and negative, of ecotourism around the globe. Along with tour operators, it also serves conservationists, park managers, government officials, lodge owners, and researchers. The society's long-term objectives include developing standards and criteria for the ecotourism industry and building an international network of professionals.

The society has produced numerous publications, including a how-to book for professionals called *Ecotourism Guidelines for Nature Tour Operators*. The twenty-page booklet makes clear the need for tour operators to educate and carefully guide travelers in order to avoid harmful environmental and cultural impacts. The guidelines cover potential problems such as polluting water sources with soaps and detergents or giving candy and trinkets to local children. "We think these are groundbreaking guidelines that provide a road map for top performance for all companies seeking to deliver ecotourism services," Wood says. "We hope these guidelines will lead to an evaluation system for wildland tours and provide a valuable reference tool for travelers, educators, and all professionals who are concerned about the impacts of tourism around the globe." More than four thousand copies of the guidelines have been distributed so far.

The society's newest publications include *The Ecolodge Sourcebook for Planners and Developers*, the third edition of *Ecotourism: An Annotated Bibliography for Planners and Managers*, and its latest *International Membership Directory*. The *Sourcebook* provides information on all aspects of developing ecolodges—typically wilderness accommodations that promote conservation education and sustainable practices to their guests through various activities. It includes a variety of architectural plans and case studies from around the world. The updated bibliography features hundreds of articles and other literature on ecotourism, some of which include results of major studies. And the directory is a first of its kind, Wood says, listing nearly eight hundred of the society's members from more than fifty countries and including more than ninety nonprofit organizations and sixty educational institutions.

In addition, the society publishes a number of fact sheets for interested tourists and tour operators. These include a fact sheet for the independent traveler, as well as others on ecotourism research, marine ecotourism, ecotourism statistics, and how to choose an ecotour operator. The how-to fact sheet includes a list of questions for travelers to ask tour operators before booking a tour to ensure that the company is legitimately involved in ecotourism. (To receive one of the fact sheets, send a stamped self-addressed envelope to the society at the address provided at the end of this chapter.) On top of all these efforts, the society also has created a Green Evaluations Program based on ecotourism consumer questionnaires to ensure member nature-tour operators are abiding by the society's guidelines.

To further education about ecotourism, the society is cosponsoring a series of special courses for businesspersons interested in developing and managing environmentally sensitive tourism businesses. With the help of George Washington University in Washington, D.C., these semester-long courses introduce participants to practical techniques to better manage and operate tourism businesses that promote meaningful travel experiences and a commitment to environmental conservation. Course titles include Ecotourism and Management, Sustainable Hotel and Resort Project Investment, Ecolodge Planning and Sustainable Design, and The Realities of Owning and Managing an Environmentally Sound Inn or Resort.

The society also sponsors other periodic one- and two-day workshops that provide people involved in the ecotourism industry with the education necessary to operate topnotch ecologically and culturally sensitive tourism efforts. At some of the workshops, participants present their own case studies for review and critique by a team of experts. "There is still a limited number of tourism professionals who can genuinely put the principles of ecotourism into action," Wood says. "These educational opportunities are designed to broaden expertise and give participants the chance to learn from experts and help set high standards within this new field." In some cases, tour operators think they are providing bona fide ecotourism opportunities, when in reality some changes in their operations are needed to make their enterprises truly environmentally sound."

In another effort to help make ecotourism a more ecologically responsible activity, Jerry Mallett, president of the Denver, Colorado, Adventure Travel Society, organized the first World Congress on Adventure Travel and Ecotourism in 1991, which was endorsed and hosted by the United Nations Environmental Program. The congress aimed at showing the world's governments how to manage and market ecotourism so the industry can improve the lives of local people while protecting their environment. The dialogue created between government and industry representatives was productive, Mallett says. Similar events have been held every year since. Nearly six hundred people from thirty-seven countries attended the 1995 congress in Tasmania, including representatives from governments, corporations, tour operations, and nongovernmental organizations.

"The ecotourism industry must be managed sustainably. We haven't found the ideal situation anywhere yet, but we're striving for it," Mallett says. "The paradox with ecotourism is that we're selling the very thing we're trying to protect. Many governments still don't quite understand the concept of ecotourism. They don't know how to define it, and they don't know how to manage it. In general, people are not yet really sure what to do with ecotourism—it's a struggle, and no one has the perfect formula. But I think it can be a management tool to bring out the highest possible value for cultural and natural resources."

As time passes, the ecotourism industry will continue to make firm its philosophy and spread its environmental aims further into the planet's massive tourism industry. Similarly, governments worldwide will continue to discover that ecotourism can benefit both their people and natural resources. In realizing that economic gains from tourism can be sustained while protecting the environment, governments will work harder to protect their natural resources—the very thing that attracts many tourists. As the saying goes, "If it pays, it stays."

FOR MORE INFORMATION
Adventure Travel Society, 6551 South Revere Parkway, Suite 160, Englewood, CO 80111; (303) 649-9016

Earthwatch, 680 Mt. Auburn Street, P.O. Box 9104, Watertown, MA 02272; (800) 776-0188

Ecotourism Society, P.O. Box 755, North Bennington, VT 05257-0755; (802) 447-2121

CHAPTER 11

STUDENT CONCERNS

Since our children will eventually inherit the earth, it's encouraging to know that youth are concerned about and contributing to the environmental movement. There are a number of different ways they can get involved. Individual activities such as writing letters to politicians and business owners can help create change, as can group efforts undertaken by youth-oriented environmental organizations. A number of these groups are working to protect the environment today and have been at it for a number of years.

For instance, in 1988 a group of students at the University of North Carolina placed a small message in the magazine published by the environmental group Greenpeace, asking if anyone would be interested in forming a student-based environmental coalition. Students from more than two hundred U.S. college campuses replied. Through campaigns, conferences, and a lot of hard work, the Student Environmental Action Coalition (SEAC, pronounced *seek*) has grown to include chapters at more than two thousand colleges and high schools in all fifty states. "With a little more outreach after that first message, the interest just flooded in," says Amit Srivastava, national organizer for the group. "At the time, there was a growing surge of interest in environmental activism on campuses. SEAC began at just the right time to harness that surge."

The organization serves as a communication link connecting different student environmental groups around the country to create a more integrated movement. SEAC gives students a sense of connection to a national environmental movement. Through this united effort, thousands of youth have translated their concern for the environment into action by sharing resources, building coalitions, and challenging the narrow mainstream definition of ecological issues. SEAC prefers a broader definition of the environmental movement—along the lines of what the *New York Times* called "a move to protect communities as well as trees." SEAC is calling for a new American environmental agenda that includes the issues of race, class, and poverty, Srivastava says, along with the traditional goals of conservation and preservation. "SEAC challenges its members to see the connections between social and environmental problems. Our environment is dying not just because humans abuse the earth, but because we abuse each other. To be strong, our movement must cross lines of gender, race, and class."

Just as biodiversity makes ecosystems more robust, a diverse mix of people involved in the environmental movement will maximize participation, support, and effectiveness. Without a broad base of support, environmental progress will be limited. For instance, communities that lack the power to defend their neighborhoods or traditional lands—as is mentioned in chapter 8—often suffer because environmental regulations are not developed or enforced for their communities. An Eskimo community at Point Hope, Alaska, lived and hunted around a radioactive waste site for thirty years, unaware that the site existed. After a high rate of cancer was detected in the community, only then did its members learn of the discarded toxic waste. They were able to force the U.S. government to remove the material and dispose of it at a facility in Nevada, but much damage to human health had already been done.

One of SEAC's biggest campaigns seeks to help stop a huge hydroelectric project in Quebec, Canada, called James Bay II. SEAC chapters have been working with labor groups to fight the project, which would displace thousands of Canadian Indians and flood wilderness areas. The Grand Chief of the Cree Indians, Mathew Coon Come, has led his tribe's effort at the local level. His goal is to maintain the self-sufficient way of life the Cree have enjoyed for five thousand years. If it were to be completed, the project would flood and destroy a watershed the size of France—an area rich in caribou and migratory waterfowl. The first phase of this project, James Bay I, already has destroyed much of the tribe's fish supply and dislocated numerous Cree settlements, causing many social problems. If the second phase of the project is completed, thirty dams and six hundred dikes would be built to block nine major rivers.

To finance the project, Quebec has been signing power contracts with U.S. utilities, Srivastava says. However, SEAC helped win a major battle when the New York state government decided not to sign a $13 billion contract to buy power from the project. The group achieved this in part by following then New York governor Mario Cuomo everywhere he traveled in America and voicing their opposition. New York will now support a campaign to reduce energy use while promoting economic development in the state. "It was clear that this victory came partly as a result of SEAC's efforts on the issue," Srivastava says. "We worked to build coalitions with labor groups to show that billions of dollars flowing out of New York would hurt the state and cost jobs. Getting the support of organized labor was a key in changing the state's mind." In doing so, SEAC realized that it had to look beyond its own environmental causes to create change. It relied on the threat to the state's economy from exported jobs and money rather than on stressing the importance of saving the Cree's way of life and their native lands. Although James Bay II has not been stopped completely, the project's developers have been forced to complete an environmental impact analysis, which has bought time for environmental groups to drum up more opposition.

In another SEAC effort, a chapter at the University of Arizona has been working with the San Carlos Apache Indian tribe to block the construction of the Columbus Telescope on Mount Graham, in Arizona. The mountain has religious significance to the Apache and is one of the most biodiverse areas in the southwestern United States. Both SEAC and the Apache believe the huge telescope complex and its related buildings and development, as well as the roads leading to and from the installation, would harm the mountain ecosystem. Through SEAC's campus network, the group is applying pressure on other universities to withdraw their support for the project. And by forcing the University of Arizona to prepare an environmental impact statement, the group hopes to block the project. Although it's only a small consolation to the San Carlos Apache, SEAC has prevented the telescope from being named after Christopher Columbus—the European who most symbolizes the white invasion of the world of North American Indians.

Over the last five years, SEAC has hosted three national conferences, drawing more than twelve thousand activists to hear speakers such as Robert Redford and Jesse Jackson. The purpose of the conferences is to continue refining the group's overall agenda and to further coordinate its members. In addition, before the Earth Summit in 1992, SEAC cofounded an international network of students from forty-seven countries to represent the voice of youth worldwide at the historic meeting of global leaders in Rio de Janeiro. The network continues its work today and now includes sixty-five countries where youths work together through SEAC to broaden the definition of the environment to include social justice issues along with conservation efforts.

Another youth organization, the Student Conservation Association (SCA), has been helping the environment since 1957 by pairing short-staffed federal agencies with student and adult volunteers who want opportunities in natural-resource

Student groups have been hard at work for the environment for years. This high-school-age crew from the Student Conservation Association tackles trail work in Yellowstone National Park. (Photograph by Carla Neasel, courtesy of Student Conservation Association, Inc. Used with permission.)

management. The nonprofit organization recruits and sends into the field more than twenty-five hundred individuals to nearly three hundred locations each year, says Scott Izzo, SCA's president. These volunteers gain valuable experience while working for agencies such as the National Park Service, U.S. Forest Service, U.S. Fish and Wildlife Service, Bureau of Land Management, and the U.S. Navy's Natural Resources Program.

SCA's High School Program provides nearly 450 students with conservation opportunities across the nation each summer. The students spend four or five weeks on a work project before being rewarded with a recreational week at the end of their stay. "Typically, a group of six to ten students works out of a tent camp in the back country, fixing their own meals and learning about living in the wild," Izzo explains. "They work together on a common project under the direction of SCA crew leaders—doing trail construction or land restoration—and also receive some education about the environment from their leaders and guest speakers. The work is hard, but it's rewarding." SCA provides the students with equipment, food, and shelter and can provide financial aid to participants who need help with travel expenses. The students take turns with camp duties, take responsibility for helping their crew complete its work, and learn from agency officials about careers in conservation. There is no tuition fee for participants.

Each year, the SCA Resource Assistant Program allows thirteen hundred more volunteers—ranging from college students to retirees—to serve with professionals from a number of different conservation agencies. The volunteers are trained as seasonal employees and can work on endangered species protection, ecological restoration, natural-resource management, and recreation management. They work for twelve-week periods throughout the year and receive one to two weeks of intensive training at the beginning of the assignment. Applications are screened by SCA and agency officials to match experience and academic skills with available openings. Opportunities include leading llama treks into the Pecos Wilderness of Santa Fe National Forest or rafting down the Colorado River in the Glen Canyon National Recreation Area to oversee riverbank campsites. Assignments in Alaska and Hawaii are the most popular. If accepted into the program, students receive funds for travel and food. Housing ranges from tent camps to apartments.

The association also operates its Conservation Career Development Program (CCDP), with offices in Los Angeles, Newark, Oakland, Seattle, and Washington, D.C. This program focuses on minorities in programs designed to increase diversification in the conservation-career field and enables high school and college students from diverse cultural backgrounds to address environmental issues in their communities. By visiting work sites, meeting with community leaders, and getting an insider's view of a conservation career, CCDP participants gain valuable career development experience. In return, they bring their diverse backgrounds to the conservation community. All SCA programs are competitive: roughly one in three applicants are chosen to participate in the program. For most positions no experience is necessary, but for some specialized positions, conservation or academic experience is helpful.

In addition to its hands-on efforts, the SCA also publishes a monthly magazine, *Earth Work*, which features one of the most comprehensive job listings in the conservation community. Job-listing information includes contact names, salary range and qualification requirements for internships, entry-level positions, directorships, and more. In each issue, more than one hundred openings with federal and state resource agencies, environmental advocacy groups, and private organizations nationwide are listed. The magazine also features the latest in conservation-career news and advice.

"We think our efforts help the environment in two ways. One is we're actually getting work done out there that otherwise wouldn't get done," Izzo says. "These agencies turn to us because they don't have the budget to hire enough people to get all their projects completed. And two, the people who participate come away with a better understanding of environmental issues and concerns, as well as some appreciation for public lands and natural resources. Hopefully, they carry this appreciation with them throughout their lives." Nearly 70 percent of the SCA's student volunteers have gone on to seek careers in conservation. To date, more than twenty-five thousand people have participated in one of the group's programs.

Another youth-related group, Earth Service Corps, is a YMCA program that provides high school students across America with the opportunity to promote environmental issues and actions. Started in 1989 in Seattle, Washington, the original YMCA Earth Service Corps (YESC) program quickly grew until students from throughout the Puget Sound area were involved. By April 1995, fifty-two Earth Service Corps programs had formed in more than twenty states. Overall, more than fifteen thousand students and five thousand adult volunteers are active in the effort. "It's a unique program," says Kate Janeway, YESC director. "There are several environmental education programs, several youth leadership programs, and several service-related learning programs. We have all of those elements rolled into one."

Unlike many other youth environmental programs, YESC students do not merely participate in environmental education or action projects under the direction of an adult. Instead, Earth Service Corps operates as a partnership between students, teachers, and the YMCA. By working together, all participants are able to take the classroom out of the school and put it into the community. Students initiate and run their own school-based service projects. Teachers, scientists, and other adults in the community act as mentors and volunteers to assist the students in their efforts. Larger regional environmental service projects such as Earth Day events or other educational programs are coordinated by YMCA staff with the help of student leaders.

As defined by the students, the environment is where you live. YESC students are able to adapt their clubs to fit their needs and create solutions to local environmental issues. Over the years the students have successfully organized their peers to create a wide variety of community service projects, Janeway says, including planting trees in inner-city neighborhoods, building community gardens and parks out of abandoned city lots, establishing school- or business-based recycling programs, going door-to-door to promote the use of mass transit systems, cleaning trash out of rivers, and teaching elementary school children about the environment through field trips and classroom presentations.

The long-term goal of the national program is to establish Earth Service Corps clubs in all 2,200 YMCAs in the United States. The U.S. program has become a model for YMCA-sponsored youth environmental programs in Hong Kong, the Philippines, Singapore, Brazil, Argentina, and Malaysia. "Earth Corps has done many things for our YMCA," says Jennifer Parker, executive director of the Metrocenter YMCA in Seattle. "It has revitalized our outreach to high school youths and has brought us into new partnerships with community leaders, corporations, and government agencies."

Of course, these are just a few environmental organizations available for students to join. Programs like these offer valuable learning opportunities for students while they complete projects and make efforts to help protect the environment. Your local library should be able to help you find other organizations at the local, state, and national level.

FOR MORE INFORMATION

Earth Corps, Program Information, 909 4th Avenue, Seattle, WA 98104; (800) 733-YESC or (206) 382-5013

Student Conservation Association, P.O. Box 550, Charlestown, NH 03603; (603) 543-1700

Student Environmental Action Coalition, P.O. Box 1168, Chapel Hill, NC 27514; (919) 967-4600

Training and career opportunities in the environmental field are more plentiful than ever before. Students in this photo study a spill-dispersion map with Dr. Roy Hann, right. (Photograph courtesy of Texas A&M University. Used with permission.)

CHAPTER 12

ENVIRONMENTAL EDUCATION AND ENVIRONMENTAL JOBS

The environmental remediation and restoration field has boomed since the late 1970s, and colleges around the world are responding with new educational programs to meet the growing job demand. America's Superfund law spawned a multimillion-dollar cleanup effort in and of itself when it was passed in 1980. Career opportunities in the environmental field today continue to expand and diversify. Whether one is interested in working in environmental law, studying endangered species, or leading environmental cleanups, the demand is there—and the work can be rewarding.

A college on Colorado's Western Slope, for instance, has created a degree program to train future environmental technicians and scientists. Mesa State College, located in the city of Grand Junction, now offers one of the nation's first bachelor's degrees in environmental restoration and waste management. Since 1990 the school also has offered a two-year associate's degree in environmental restoration engineering technology. "But there was a strong need for a four-year program to help students gain a higher level of expertise," explains Dr. Karl Topper, who heads the program and holds a doctorate degree in soil and analytical chemistry. "And there are more opportunities for students with bachelor of science degrees. For instance, government regulatory agencies want their new hires to have these degrees."

Both degree programs are interdisciplinary—requiring students to take classes in geology, math, chemistry, writing, and biology along with their core environmental curriculum. In addition, students must complete ten laboratory classes. Each program prepares students to solve difficult problems both in the cleanup of sites contaminated by toxic waste as well as in managing wastes that are continuously generated by industry. "We're hoping to train people who can do this type of work more efficiently and effectively, as well as bring new ideas to the field," Topper says. "The key is industry and the government being receptive to introducing these ideas into their systems."

To bridge the gap between academics and industry needs, students perform actual fieldwork such as sampling wells and using various types of instruments to perform tests. The college also is trying to establish a training facility to allow for more hands-on exercises that simulate real-world activities. "We already have the land and a building but need money to remodel the facility and buy equipment," Dr. Topper says. "We're encouraging new partnerships with industry and governmental agencies."

When the program started, AT&T made a significant donation of computer equipment. That equipment is now being used by students to develop software aimed at assessing risks at various contaminated sites. Thanks to a grant from the U.S. Department of Energy, the students also are rewriting some of the department's existing software, adding capabilities and new applications. In addition, Scientific Measurement Systems, a Grand Junction company, donated $100,000 worth of instrumentation to help students in the Mesa State program learn how to analyze various environmental samples such as soil, water, and heavy metals. "We have plenty of students interested in the program," Topper says, "but we can't keep moving forward without the continuing help and support of both industry and government sources."

In that vein, the National Science Foundation awarded Mesa State a $400,000 grant in 1995 to create an environmental education program for Native Americans. "This project should have a positive impact on the ability of Native Americans to take care of their own environmental problems," Topper says. "Unfortunately, much of their reservation land has been damaged by past industry- and government-supported projects, not to mention the other problems facing many reservation communities, such as waste management and water conservation. However, Native Americans seldom have the required education and training to address these issues, relying instead on nontribal consultants to provide technological support."

Native Americans have been victimized by a number of environmental injustices over the years. Some of their land has been desecrated by illegal dumping in the form of everyday trash or industrial wastes, and irresponsible mining practices also have contaminated water and soil. By employing their own environmental professionals, Native Americans can gain greater control over their local environments and help improve health and economic conditions. Furthermore, Native American environmental professionals won't be inhibited by communication barriers or a lack of understanding of tribal culture and religion. Not surprisingly, the National Science Foundation grants are aimed at colleges that serve a number of Native American students. Navajo Community College in Shiprock, New Mexico, which is located on a reservation, also is supported by the foundation. "Eventually, we would like to transfer the results of this project to other appropriate colleges around the country," Topper explains.

Anyone in pursuit of job security in the environmental job world could take a look at the oil-spill-response business. As is mentioned in chapter 5 on oil-spill remedies, every year there are about ten thousand spills on America's waterways alone. And if you count chemical spills, the numbers triple. Dr. Roy Hann Jr. has been leading a fight against these disasters for more than twenty years. Applying his experience with tanker, pipeline, and facility spills, Hann helped start the world's first oil-spill cleanup program at Texas A & M University in 1974. His goal is to teach students and industry how to prevent oil spills and ways to minimize the damage when spills do occur. He has prepared and presented oil-spill courses for Brazil, Chile, India, and the United Nations, and he has worked with the U.S. Navy and Air Force on cleanup projects. "The job market for our environmental engineering graduates is booming," he says. "Since the *Exxon Valdez* spill in Alaska, our enrollment has doubled. We even have a backlog of students trying to get into the program." Currently the program is limited to forty or fifty new students each year.

Like most problems, it's easier to prevent a spill than to clean one up. In the long run, prevention saves both time and money as well as the environment. "Most companies know what should be done but don't actually do it for budgetary or other reasons," Hann says. That's why he stresses a system called the "oil-spill-prevention cycle"—a thought process that starts with ship and facility design and considers all aspects of the shipping business, including construction, operation, maintenance, training, supervision, and enforcement.

Since Texas A & M's campus at College Station is landlocked, only research work and basic training can be done there. However, Hann's graduate students travel to the Texas coast each year to conduct realistic oil-spill-response exercises under the watchful eyes of the U.S. Coast Guard and the Texas General Land Office—the state's oil-spill agency. "We conduct a drill for the students that is comparable to an oil spill one-third the size of the *Exxon Valdez*—simulated, of course," explains Hann. "We take them down to the oil terminal so they see storage tanks and tankers coming in. Then we create a hypothetical event that they have to deal with. They put a little bit of oil in the equipment just to get a feel for it, but for the most part it's just a management activity. The students set up a command center, operate lots of computer programs, keep updated situation boards, and call hypothetical contractors. Then we have another group that simulates the response. We actually call state agencies and report the situation, identifying it as a drill, and the agencies respond and participate, too. We try to make it as realistic as possible." The drill starts at 6 A.M. on a Saturday and it runs straight through until noon on Sunday.

Many of the program's graduates find jobs with oil companies, environmental engineering firms, or in government, working on both prevention and cleanup issues. Hann's graduates earn either a master's or doctoral degree in civil engineering with an emphasis on environmental engineering. "I've seen some of my graduates go on to participate in government committees and help make substantial improvements in our country's laws," he says. "Overall, people are better prepared for oil and chemical spills than they used to be. They have better equipment, and they're required to hold emergency response drills. We really have seen some major improvements in the field concerning our ability to deal with the problem."

Hann secured a multimillion-dollar service contract from the U.S. Air Force in 1995 to help clean up some of its contaminated sites in America. The contract is primarily for professionals, but the Texas A & M program's graduate students also will benefit from the experience. The contract is for the cleanup of hazardous materials, prevention consultation, and restoration. "I went on a sabbatical with the Navy for six months at one point, helping them with some of their oil-spill problems on the West Coast. My involvement began with restoration projects and moved on to include prevention management as they caught up on their backlog of problems," Hann says. "Eventually, I reviewed their oil-spill-response courses and also reviewed their whole response capability in the San Francisco Bay Area. I expect the project with the Air Force to be similar."

Hann is one of only a few engineering professors in the country who have been devoted to spill-response management and technology over the last twenty years. He responded to the world's worst oil spill, which occurred in 1991 in the Persian Gulf during the Persian Gulf War, as well as to the largest tanker spill to hit a shoreline, when the Amoco ship *Cadiz* broke up in March 1978. That massive spill contaminated nearly two hundred miles of France's Brittany Coast and was six times larger than the *Exxon Valdez* spill in Alaska. "The *Cadiz* spilled 220,000 tons of light Arabian and Iranian crude. The *Exxon Valdez*, by comparison, lost only 38,500 tons of oil," Hann explains. "The *Cadiz* spill occurred due to a design flaw in the rudder—the rudder blew up under pressure and ruptured the tanker. It couldn't be fixed, and the entire load was lost." France brought in its military to supply the manpower for the cleanup, and Hann visited the spill site four times over the next year to monitor the military's progress. He found a number of marshes along the coast where oil had collected, and it remained in those marshes five years after the spill.

On the Alaskan coasts near the site of the *Valdez* accident, the beaches look clean and healthy from a distance of twenty feet or more. "But up close, you can see that the marine life is all one size," Hann says. "It doesn't have visible generations because the older ones were heavily impacted." As we see in chapter 5, "Remedies for Oil Spills and Their Ills," a large-scale oil slick can affect life at all levels—from seabirds and bears to fish and lobsters. The cumulative results of an oil spill can be devastating to numerous species and can last for decades.

That's why programs like the one found at Texas A & M University are so important. By training tomorrow's environmental professionals, these programs can lead industry toward safer environmental practices and help respond to crises when they occur.

FOR MORE INFORMATION

Mesa State College, Department of Physical and Environmental Sciences, Grand Junction, CO 81501; (970) 248-1654

Texas A & M University, Department of Civil Engineering, College Station, TX 77843-3136; (409) 845-3011

NATURAL RESOURCES

Whose land is it, anyway? The Trust for Public Land organization helped protect eight community gardens and pocket parks in Boston's South End. (Photograph by Susan Lapides, courtesy of the Trust for Public Land. Used with permission.)

CHAPTER 13

SAVING LAND FOR PRESENT AND FUTURE GENERATIONS

"This land is your land, this land is my land." Simple lyrics to a well-known song, but lyrics that contain the crux of a difficult environmental problem. What land belongs in the hands of private owners, and what land should remain open to the public for all to enjoy?

No easy answers here. Some people argue that there already is plenty of publicly owned land available for ordinary citizens to enjoy in the form of parks, wilderness areas, and city- or county-owned open space. Enough is enough, they say. However, on the other side of the issue are environmentalists who claim that U.S. metropolitan areas continue to gobble up land as these cities expand into nearby countryside without providing needed open space. Subdivisions and shopping centers as well as numerous other forms of commerce and housing can seemingly spread out forever in many U.S. cities, they argue, and humanity has a basic need for open natural settings—be they in a metropolitan area or several hours away in wilderness. Both settings can provide a respite from the chaos of modern society and can help bring humans back in touch with the natural world. This in turn can lead to a balanced planning approach in terms of future uses of our cumulative land resources.

Land is one of the world's most important natural resources. It provides humanity with food and shelter, among numerous other things contributing to human survival. However, it also provides for the very existence of the plant and animal worlds. This section of the book addresses the crucial natural resources of the earth and describes efforts currently under way to protect and conserve them. Subsequent chapters in this section discuss other resources, including air, water, and forests. All play a critical role in maintaining the overall health of the world's environment, and all of these resources are at some risk today.

Concerning land, one U.S. nonprofit effort has pursued the acquisition of various parcels for public use on a national scale since 1972. The Trust for Public Land aims at protecting land important to America's heritage across the continent—be it expansive wilderness or a city lot set aside for a park or garden. The private organization operates twelve offices around the country and has completed more than one thousand projects, protecting nearly one million acres of land in forty-three states and Canada. The fair market value of the land tops $1 billion. "The Trust for Public Land is a conservation, real-estate organization," explains its president, Martin Rosen. "We work in the marketplace on transactions that result in land for public use—and many times have to solve complex problems associated with land deals. Basically, we see that the public is represented at the negotiating table."

The trust works closely with both community groups and government agencies to acquire and preserve land to serve human needs. It is the only national conservation group specifically established to conserve land for public use and open space, Rosen said. When pieces of land are deemed valuable to the public, the trust works to find stewards for the property, which can include government agencies or nonprofit organizations. Because of the trust's nonprofit status, landowners can donate their land or sell it to the trust at below market value and receive an income tax reduction.

In many projects the trust helps grassroots community groups acquire key parcels of land. Some projects involve preserving river corridors and forests, and adjoining lands may be used for development such as housing. In Grand Junction, Colorado, for example, an unsightly junkyard beside the Colorado River was transformed into a riverfront greenbelt. In New York City, an asphalt traffic triangle was reborn as a park.

The trust protects land in both rural and metropolitan settings, but much of its work is accomplished in city neighborhoods underserved by parks and open space. Its Green Cities Initiative targets twelve large U.S. cities where the trust believes opportunities, leadership, and community commitment will lend support to comprehensive land acquisition and park-improvement programs. The five-year project includes the metropolitan areas of Boston, Seattle, Atlanta, Cleveland, Baltimore, New York, Los Angeles, and San Francisco, among others. The initiative aims at increasing public awareness of the vital role parks and open space play in the quality of urban life. The trust plans to generate funding to create, improve, and maintain urban parks in these cities as well as to protect public open spaces that preserve and celebrate a city's unique heritage. To this end, the trust hopes to eventually acquire up to 250 properties that meet high-priority needs in these cities and to raise $2.5 billion in public funding for urban open spaces.

"We see our unique role as connecting land with people—and it is in urban areas that most of America's people live," Rosen says. "Having a natural place to go is important to people. We see this effort as a way to bring more people into the environmental movement. A city lot full of vegetables and flowers offers a touch of nature for residents that can mean more to them than a piece of land a hundred times larger in a remote setting. And because the projects typically are accomplished through community involvement, they can help instill in the population a respect and sense of responsibility for the environment."

Another unique land-acquisition group, Rails-to-Trails Conservancy, draws its resources from America's great heritage in railroads. In the late 1800s, the United States started building a huge interconnected system of railroad lines. By 1916 that system had become the largest in the world, with more than 250,000 miles of track. But as transportation methods changed during the twentieth century, thousands of miles of these railroad tracks began falling out of use. Today, more than three thousand miles of track are abandoned each year.

In an effort to conserve these corridors for public use, Rails-to-Trails is working to convert abandoned railroad lines into multiuse trails. Founded in 1986, the private nonprofit group already has celebrated the opening of the five hundredth rail-trail—an eleven-mile stretch in Massachusetts called the Minuteman Trail, which runs through the historic town of Lexington and follows part of the route Paul Revere rode in April 1776. The entire rail-trail network the conservancy helped create has passed the seven-thousand-mile mark, says Steve Emmett-Mattox, the group's community affairs coordinator.

Rail-trails can be found in forty-eight states—with Michigan leading the way with eighty-seven trails. Another one thousand projects are under way across the country that will add more than twelve thousand miles of trails. As former railroad beds, the trails are flat and well graded and thus amenable to a number of uses, from bicycling and walking to horseback riding and cross-country skiing. For those confined to wheelchairs, auto-free paved trails provide a much-needed resource. The trails also provide plenty of recreation on very little land. A mere two hundred acres of former railroad right-of-way can create a park and trail more than twenty miles long.

Creating rail-trails became viable after Congress passed the 1983 National Trails Act, which kept railroad corridors intact for future use as trails and open space, or even for a return to railroad tracks. "Our goal is to convert as many of these abandoned railroad corridors into trails as possible and create an interconnected network of trails throughout the country," Emmett-Mattox says. "Right now, you could ride on a trail from Pittsburgh all the way to our headquarters in Washington, D.C."

Much like the Trust for Public Land, the conservancy acts as a resource to help various volunteer groups, private foundations, and park agencies create new trails. The organization assists these local groups in following proper legal procedures and seeing the sometimes lengthy acquisition process through to completion. "Once created, rail-trails can serve both recreational users and daily commuters—linking homes, businesses, and parks. They provide useful routes away from the world of motorized vehicles," Emmett-Mattox says. "Many times, these trails are the last section of open land in urban areas. They are a swath of green that presents an opportunity to provide a natural setting and environment."

In a smaller land acquisition and preservation effort in California, a wilderness council composed of ten Native American tribes has been working since 1986 to create America's first intertribal park along the state's northern coast that will be open to the public. The 3,800-acre tract of land, located in northern Mendocino County, once flourished with dense fir forests, tranquil oak meadows, and towering stands of fog-shrouded redwood trees. The land also once supported the largest concentration of Native Americans in North America. Studies of this part of California show that the native Sinkyone people (pronounced *sin-kee-own*) inhabited the land for ten thousand years, said Hawk Rosales, coordinator of the InterTribal Sinkyone Wilderness Council.

That is, until the mid-1800s. For twenty harsh years, white settlers and the U.S. military massacred many of the native Sinkyone people and chased the remainder off their land so that mining and logging activities could commence. Sinkyone children even were sold into slavery in San Francisco. These human rights abuses led to an exodus by the Sinkyones. In a few years, land that had supported Native America's most densely populated region saw its native people disappear. Once the land had been cleared of its longtime human inhabitants, logging efforts soon began clearing the land of another native inhabitant.

Over the next century, those efforts ended up cutting down most of the old-growth redwood trees on the land—in essence destroying the lush temperate rain forest. Today, only 2 percent of the area's original old-growth forest remains standing. When a redwood forest such as this is felled for timber production, centuries worth of growth disappear and the land takes on an entirely different character. Wildlife habitat is lost, forcing birds and animals to live elsewhere. Similarly, many native plants that thrived in the temperate rain-forest ecosystem can't survive on land devoid of trees. Without the trees and plants, erosion becomes a problem as unstable soil is simply washed away. The natural balance the land supported before logging is destroyed. "We're now looking at a degraded watershed and ecosystem because of overlogging," Rosales says. "Much of it is devastated from clear-cutting and subsequent erosion. Since 1986, we have proposed to bring communities together and use both native and modern approaches to try and restore the land."

A lawsuit in 1986 stopped the Georgia Pacific logging company's plans to clear-cut the Sally Bell Grove after the California Department of Forestry had approved the sale. The ninety-acre grove of old-growth redwoods is named after one of the Sinkyone massacre survivors, who watched as her parents were slaughtered by white vigilantes. Later, a California appeals court ruled that in approving the timber sale, the department of forestry had violated the state's environmental laws. The state government—through its Coastal Conservancy Board—then ended up buying 7,100 Sinkyone acres from Georgia Pacific in 1986 for $1.1 million. Roughly half of the land, that nearest the coast and in better condition, was added to the existing Sinkyone Wilderness State Park. The remaining 3,900 acres, called the Sinkyone Uplands Parcel, was designated to be resold by the conservancy board in 1995. The timber industry hoped it could purchase the land to conduct commercial logging operations, Rosales says, but the wilderness council set its sights on creating a unique park in tribute to its ancestors.

The Coastal Conservancy Board ended up siding with the wilderness council, agreeing to sell the land to the group. "Our goal was to acquire the land, and we now have a chance to restore it to its original balance with Native American uses and stewardship," says Rosales, a Chiricahua Apache. "Our land-purchase fund has just over $100,000, but by the middle of 1997 we must raise a total of $1.4 million. We believe we have developed proper long-term land-management goals but are struggling to meet the financial demands of the land purchase. If we cannot raise enough, the next buyer will be a logging company." The group hopes to raise the funds through grants and private donations, or even loans that could be paid back through revenues eventually generated by the park.

The wilderness council already is leading efforts to restore the land. By bringing together university researchers and other professionals with Native American community members, restoration projects are combining traditional land-management methods with modern scientific approaches. Work is under way to remove logging roads and recontour parts of the land as well as to restore some of the streams that once teemed with salmon. Various foundations and government sources are providing funding for these restoration efforts. Other projects include maintaining a nursery for native trees and plants, conducting tree-planting and brush-removal projects, and preserving cultural remains from the past. "The land contains many important cultural resources and sites sacred to Indian people of this region," Rosales says. "Many Native Americans are still strongly connected to this land, which they use for ceremonial and food-gathering purposes. We want the park to be a living model of Native American land stewardship that Indian people practiced here for thousands of years."

Eventually the council will open the park to the public and manage it with local Native Americans. The area contains many sacred archaeological sites that remain culturally significant to the descendants of the Sinkyone people and would be of interest to all visitors, Rosales says. In addition, four small village sites will be incorporated into the park—all built with traditional materials in traditional ways. The sites will be used for retreats, cultural activities, and other ceremonies. A documentary film, *The Run to Save Sinkyone,* has been produced by the Sinkyone Wilderness Council about the fight to save the land and future plans for the park. Three years in the making, the film won an award in 1995 at Robert Redford's Sundance Film Festival in Utah and is available for thirty dollars a copy from the council.

"The Indian people who descended from the Sinkyones are now repairing the damage to their homelands," Rosales says. "The land has been severely degraded by industrial logging practices, but we believe our project will create a living model that incorporates traditional Indian land-use practices with modern restoration efforts and restores the ancient redwood forest ecosystem. Our dream is that all people will one day enjoy it through a California Indian perspective. By returning Sinkyone land to Indian hands, we can establish a landmark environmental and cultural restoration effort—one with both practical and spiritual benefits for the land and the people."

Conflicts over land-use issues are now commonplace as supporters of land privatization square off against backers of public lands and prodevelopment forces battle with conservationists. If you support the concept of land being set aside for public use as well as the conservation of wildlife habitat and sensitive ecosystems, join in these decisionmaking efforts and discussions. And if you believe in the public-land concept but don't care to get personally involved, supporting the efforts in this chapter also help make a difference. If more parcels of land are opened for public use and enjoyment, more people will have the opportunity to gain a better understanding of the natural world and, consequently, the need to protect it.

FOR MORE INFORMATION

InterTribal Sinkyone Wilderness Council, 190 Ford Road, #333, Ukiah, CA 95482; (707) 463-6745

Rails-to-Trails Conservancy, 1400 16th Street, NW, Suite 300, Washington, DC 20036; (202) 797-5400

The Trust for Public Land, 116 New Montgomery Street, Fourth Floor, San Francisco, CA 94105; (800) 714-LAND

An abundance of safe, clean water is a necessity for all life.
(Photograph by Kevin Graham.)

CHAPTER 14

SOLVING WATER PROBLEMS

Water: Essential to existence. The basic mechanics of the planet depend on it. Industry around the world relies on it in order to produce many of the products we depend on for daily living. Humanity's very future is dependent on the availability of clean drinking water. But it must be purified after its many uses in modern industrial society, for it is harmful if polluted. New sources of the precious liquid are always welcome, of course, although not easy to come by.

Water is in increasingly short supply as the earth's human population soars. Obviously, more people on the planet means more water must be consumed to sustain humanity as a whole. But in addition, higher human populations create problems in keeping the world's freshwater supplies clean and in ensuring that they are not contaminated by the additional wastes produced by burgeoning population and industrial bases. Fortunately, numerous scientists, nonprofit groups, and government agencies are tackling both the supply and the end-use issues of the critical water equation.

One project is focused on a new form of "pyramid power." It's hard to imagine a pile of rocks providing a steady source of fresh water, but an ancient technology—reaching back 2,500 years—is being researched by a Seattle engineer for its water-production potential in modern society. Jose Vila, who holds a doctoral degree in civil engineering, first came across the phenomenon while studying road-buckling problems in the Southwest. Roads built in this arid region were mysteriously damaged by moisture. Vila's research revealed that the rocks, or aggregate, used in the bases for the asphalt roads were condensing water moisture from the air. When the water then dripped into and expanded clay soils beneath the aggregate, the roads buckled. Engineers working on the problem dubbed the occurrence "hydrogenesis."

Since then, Vila and two partners have started a company of the same name—Hydrogenesis Inc., based in Mill Creek, Washington. Eventually they hope to build conical pyramids of loosely piled rock to capture large amounts of water vapor from the air and produce fresh water. Although the idea was used on a small scale in European farming practices for hundreds of years, Hydrogenesis's efforts are the first aimed at producing water for human consumption on a large scale from these so-called aerial wells.

It is believed the ancient Greeks originated the idea of making use of the condensation created by the daily heating and cooling of the air by the sun. At the site of an ancient Greek city near the Black Sea in the Ukraine, the remains of an amazing water-supply system were unearthed about a century ago, Vila says. A dozen pyramids, each about forty feet tall and composed of pieces of limestone, were connected by a series of clay pipes. Some archaeologists estimated the system could have produced up to fourteen thousand gallons of water a day. Although Vila thinks that number is high, he said it proves that the system has worked in the past.

In Greece today, rocks are piled around the bases of trees in a primitive attempt to provide them with some moisture. And desert mice are known to make small piles of rocks to produce drinking water. Although many aspects of the hydrogenesis phenomenon remain to be studied, Vila has completed an initial research phase and believes forty-foot-tall pyramids composed of four-inch-diameter porous rocks—such as chunks of old cinder blocks—could produce five hundred gallons of water a day per pyramid. The aerial wells will work best in arid regions with big temperature changes between day and night, he explains, and water production will vary from season to season. For instance, a pyramid's inner portions will stay cool long into spring and summer, causing increased condensation when warm air passes over and through the cool rocks. Water would then drip down to a concrete basin and into a drain for collection.

"These pyramids would be inexpensive to build, require little equipment and effort, and be harmless to the environment," Vila says. "They could be a very valuable source of water in areas where there is little inherent moisture. And no power or pumps or maintenance are required. They run on solar energy, so air pollution and waste products are not a concern." Aerial wells could provide a secondary source of drinking water in arid regions of the world, especially Third World countries, he adds. The process also could be developed to supply steady amounts of water for irrigation.

Hydrogenesis Inc. has used a National Science Foundation grant, obtained through the foundation's Small Business Innovative Research Program, to conduct initial research. The company has developed a computer program that allows it to predict aerial-well water yields throughout the year at any location and climate. The company is now looking for funding to build a prototype pyramid and measure the water amounts produced. Roughly a third of the earth's land mass is arid and could potentially benefit from Hydrogenesis's efforts.

"While aerial wells appear to be simple piles of rock involving a low-tech analysis for their design, this is not the case. What we have found, instead, is that the design of an aerial well requires an involved thermodynamic analysis," Vila says. "The technical analysis has progressed to a point where we can evaluate the climatic environment of a potential site and predict water yields through computer modeling techniques. The water yields we have obtained for a theoretical aerial well in the Colorado-Utah high desert area are very encouraging. These estimates support our conviction that these wells will prove to be a meaningful source of water in arid regions. Eventually, this could help various parts of the world recover from drought conditions. The process has tremendous potential."

Of course, producing greater supplies of fresh water is not the only water-related dilemma faced by modern society. There also is the problem of what to do with water once it has been used and disappears down the sink, toilet, or shower drain. By using wetlands—often called "nature's kidneys"—as filters, Horry County in South Carolina is processing wastewater and saving energy while leaving the land unharmed. Wetlands are a fragile ecosystem that features wet and spongy soil. These areas act like kidneys by separating waste materials from wastewater, allowing the waste to be absorbed into the soil as nutrients and eventually releasing purified water into nearby creeks, streams, and rivers.

In Horry County—the state's fastest growing county, with the city of Myrtle Beach as its hub—a sewage-disposal system is discharging 650,000 gallons of wastewater a day into wetlands, and the natural settings are thriving. "Fifty percent of the county is wetlands, so treating our wastewater this way became a natural alternative to consider compared to a typical treatment facility," says Larry Schwartz, an environmental planner with the Grand Strand Water and Sewer Authority, which handles wastewater treatment for the county. "We're just using the natural filtering ability of the land to renovate wastewater."

The wastewater is distributed evenly across three Carolina bays, which are egg-shaped natural depressions unique to coastal regions of the Southeast. These bays are the only places Venus's-flytrap—the infamous carnivorous plant—can be found in the United States. Filled with peat and shrubs, the bays act as buffers between the uplands and the region's fragile black-water rivers, so named because they are dark-colored, slow-moving, and hold small amounts of dissolved oxygen.

Only one bay at a time receives wastewater. With all three bays used alternately, up to 2.5 million gallons of wastewater a day could eventually be treated as the county grows over the next twenty years, Schwartz says. A series of boardwalks crisscrosses the bays to support distribution pipes carrying the wastewater after it is treated initially at the county's wastewater treatment facility. Two-inch-diameter holes spaced every fifteen to twenty feet allow the water to splash onto rocks and disperse evenly across the bay. The wastewater has been treated to secondary levels before it enters the bay, meaning 85 percent of all organics and wastes have been removed. Normally, to finish cleaning the water using man-made treatment systems, large amounts of energy are required.

"But in this case we're using energy from the sun, which in essence powers the wetlands," Schwartz explains. "That's the beauty of it. The system is cost-effective and energy-efficient." It's as simple as it sounds: Wastewater pours into the wetlands, where nature removes the secondary wastes and uses them as nutrients before dispersing clean water into nearby streams and rivers.

A dozen government agencies at both the state and national levels are involved in the project. Two biologists work full-time testing the quality of the water leaving the bays and studying the natural habitat of the wastewater wetlands. "Our goal is to maintain the value of the natural communities in the bays and manage them so any changes are minimized," Schwartz adds, "and in the process achieve advanced wastewater treatment."

The project has generated plenty of interest from other states, including some that plan to construct their own wetlands for similar purposes. Schwartz gives at least one tour of the county's wastewater system a month. "It's a new way to do things," he says. "If wetlands are selected and managed properly, there is no reason not to use them."

Although the wetlands of Horry County are helping maintain the integrity of the area's rivers, many American rivers—crucial to much of the nation's drinking water supplies—face a harsh future due to overreliance on their water and abuse of their ecosystems. They are the veins and arteries that bring life to land and help sustain its inhabitants. Major river threats include dam building, overdraining for agriculture, loss of riverside habitat to development, and run-off pollution from pesticides, fertilizers, road salt, and numerous other contaminants.

One nonprofit group of more than fifteen thousand members is working to protect and restore this country's river systems. Appropriately named American Rivers, the effort has protected twenty thousand miles of rivers and five million acres of riverside land since its founding in 1973. "We are the principal river conservation organization in North America," says Randy Showstack, the group's director of communications. "We work on a variety of issues ranging from protecting wild and scenic rivers to restoring urban rivers and tracking various water issues."

For example, American Rivers played a role in a 1995 success story in California. An irrigation district that had planned to construct a four-hundred-foot-high dam on the Clavey River near Yosemite National Park withdrew its plans. The district realized that the benefits gained from damming the river were not worth destroying the ecosystems supported by the free-flowing Clavey. Dams create artificial variances in the flow of rivers that can disrupt the hatching of fish and the various habitats relied on by wildlife. These variances, created when water is alternately released and impounded, also can destroy wetlands and other ecosystems located alongside rivers.

The Clavey River had been part of American Rivers' annual list of the ten most endangered North American rivers, which was aimed at informing the public about the numerous and ongoing threats to various rivers and streams. The organization worked directly with the irrigation district along with several local environmental groups to help stop the dam project. American Rivers provided the local environmental groups with advice, legal support, and financial backing to save that section of the Clavey.

American Rivers is dedicated to educating the public about the importance of rivers to American society, Showstack says, because the overall ecological health of rivers is declining at an alarming rate due to the increasing effects of human development. More industry, shopping malls, and highways, for example, mean more potential damage to rivers through added water needs and pollution. Already, 34 percent of North America's fish species are classified as rare to extinct, along with 65 percent of its crayfish species. Flowing rivers connect the mountains to the sea and nourish the various ecosystems they pass through along the way. They also serve as corridors for migratory birds and fish and are home to many other plants and animals.

"Lakes are part of river systems, and reservoirs are simply rivers backed up. Also, underground water supplies are hydrologically connected to rivers," Showstack adds. "Without healthy and clean rivers, we would have much less fresh water available for fish and wildlife or human use." The organization's efforts include lobbying, litigation, public education, cooperative negotiations with government and industry, and support of hundreds of grassroots river-saving groups. Of the nation's 3.5 million miles of rivers, 600,000 miles—or about 17 percent—already have been dammed by more than sixty-eight thousand projects. "We do use most of our dams for electricity, flood control, and water supply, but we don't need to destroy our last few wild rivers to get more of those benefits.

One of those rivers is the Clark's Fork of the Yellowstone River in Montana, which has attracted the attention of American Rivers in recent years. The group is fighting a proposed gold mine, owned entirely by Canadian companies, that would be located near the river in Montana about two miles from Yellowstone National Park. To contain the 5.5 million tons of acid waste generated by the huge mine, a ninety-foot-high dam would be needed to create a storage reservoir. Run-off from mining operations throughout the United States has damaged more than twelve thousand miles of American rivers, Showstack says, and this mine and its related dam present a risk to one of America's most popular parks, its rivers, and the surrounding region. Again, American Rivers is helping local environmental groups that are fighting against the project, as well as raising public awareness across the nation and lobbying in Washington, D.C., against federal mining laws that continue to allow rivers to be harmed.

The conflict boils down to an argument for or against development at the expense of the environment. Backers of the mine say the dam and the pollution related to the effort are worth the added economic benefits provided by the mine. American Rivers and other environmental groups involved in the dispute claim there is only one Yellowstone River and Yellowstone National Park but thousands of gold mines, as well as plenty of other areas still open to new mining operations. Why risk destroying another scenic river and a historic national park for another gold mine?

American Rivers is also working to increase the number of river miles included in the National Wild and Scenic Rivers System established by Congress in 1968. The system preserves portions of exceptionally beautiful rivers and now includes more than 150 sections of rivers that together cover about 10,700 miles. This represents less than one-third of 1 percent of the nation's 3.5 million river miles. Only nineteen states have studied their own rivers to learn which are worthy of saving from development. "We're trying to alert the public and decision-makers that our rivers—so valuable to public health, ecosystem protection, and recreational use—are not in good shape," Showstack says. "There is an urgent need to restore and protect them. We think North America's rivers should be dealt a much kinder hand."

By protecting rivers and their related ecosystems as well as searching for new sources of water, even in the form of rock pyramids, we can assure ourselves of a substantial supply of this crucial substance in the future while still protecting important fish and wildlife habitat. And it's really not a matter of wanting to do it but of *needing* to do it. If humanity does not have enough clean drinking water to support life in the future, nothing else will matter.

FOR MORE INFORMATION

American Rivers, 1025 Vermont Avenue NW, Suite 720, Washington, DC 20005; (202) 547-6900

Grand Strand Water and Sewer Authority/Wetlands, P.O. Box 1537, Conway, SC 29526; (803) 347-4641

Hydrogenesis/Aerial Wells, 14814 18th Court SE, Mill Creek, WA 98012; (206) 337-3606

Treating the earth, our home, with love and respect is a reflection of the gratitude we feel for its beauty and bounty. (Photograph by Kevin Graham.)

CHAPTER 15

LOOKING AT NATURE'S BIG PICTURE

"We abuse land because we regard it as a commodity belonging to us," naturalist Aldo Leopold wrote. "When we see land as a community to which we belong, we may begin to use it with love and respect."

Treating land as a commodity was not a practice engaged in by Native Americans before the arrival of white settlers. Rather, it is a bad habit modern society has adopted. Native Americans believed their connection to the land was a spiritual one. They respected the land for what it provided—the means for their very existence. Today, many indigenous cultures around the world continue to honor a spiritual connection to the land they occupy and therefore treat it with reverence. Now, new efforts are under way to introduce this way of thinking into modern American society.

By way of love and respect for the land, Peter Berg wants people to think about the place in which they live, understand its various natural systems, and consider how their actions can adversely affect those systems. To reach that goal, Berg started the Planet Drum Foundation, a San Francisco–based organization that works to spread the concept of *bioregionalism*. "A bioregion is usually outlined by a watershed and encompasses all the activity that takes place in that area—which involves soils, plants, animals, land forms, and human activity," Berg explains. "We want people to think about how they can live in their bioregion sustainably without harming the planet. We want to put them back in nature."

Bioregionalism involves this effort of understanding the natural systems operating in the area where one lives. Through this understanding, Berg hopes people will realize the importance these systems play in the overall environment and in turn work to preserve and protect them. He developed the bioregional concept in the early 1970s after studying the ways indigenous peoples respected and carefully dealt with the land where they lived. He sat down with ecologists and activists to figure out how to put modern society back in nature. He aimed at helping people revere the geographic terrain around them, something similar to the thinking of many indigenous cultures isolated from modern society.

Too often, things like new shopping malls, subdivisions, and many forms of industry take precedence over the land they continue to swallow up. Berg's goal is to try and help people move away from the development-at-all-costs viewpoint and think about the land they occupy in a new way. Critics, of course, contend that development is a positive part of modern society and spells progress. There's already plenty of wilderness and open space, they argue, and development should claim much of the rest.

Although some development is bound to occur over time, a possible solution is to allow for development while also protecting natural systems. Housing developments could be limited in size, for instance, if this results in the conservation of river or wetland habitats. In other cases, however, development simply should be off-limits. For example, the U.S. Bureau of Reclamation proposed constructing two dams in the Grand Canyon in the 1960s. Only a hard-fought battle by environmental groups such as the Sierra Club led the government to back off this idea. Today, it's hard to imagine many people arguing that water should fill parts of the Grand Canyon.

There are now more than 250 groups working to follow Berg's ideas in bioregions across the country. Most of the groups produce publications to promote their efforts and meet regularly to tackle various environmental projects. "Being bioregional is a way of taking care of the parts in an effort to maintain the whole," Berg says. "Maintaining the viability of the place you live is a contribution to taking care of the planet as a whole. Being bioregional is simply a way to 'act local, think global.'" He cited a good example of bioregional thinking from Tucson, Arizona, where a brochure informed people how best to live in the Sonoran Desert. "Learn to love cactus and sand," it stated bluntly. "Forget about lawns."

Bioregional groups work to maintain various natural systems such as forests, rivers, or wetlands in addition to finding ways to sustainably meet basic human needs. For instance, finding sustainable ways to meet human needs can mean obtaining food and water from local sources instead of relying on produce and water shipped from hundreds or thousands of miles away. Group efforts include restoration projects like reintroducing salmon to their native streams and rivers and monitoring timber sales, grazing leases, and road-building. The groups also promote the use of renewable energy sources and support the overall effort of what Berg terms "reinhabitation," which means appreciating the place in which you live and ensuring the long-term health of the local environment.

"Connecting people with the overall life of the place where they live is to make them conscious of their relationship to other natural entities that they depend on for life," Berg says. "So it makes them want to live sustainably within that place. This is contrary to the frontier mentality of American industrial society, but it is necessary if we are going to inhabit the planet in the long run."

Another effort based on land issues also battles America's lingering frontier mentality, which is bent on pursuing all forms of development no matter the environmental costs. The nonprofit publication *Wild Earth* calls itself an uncompromising quarterly journal that speaks for life, biodiversity, and untrammeled wild nature. Published by the Vermont-based Cenozoic Society, *Wild Earth* features wilderness proposals from throughout North America as well as regular sections on biodiversity, land ethics, overpopulation, and tactics for conservation activists.

"*Wild Earth* is an independent voice for what we call the new conservation movement," says its editor, John Davis. "We work with organizations throughout North America to promote wildlands restoration and protection. Our aim is to present excellent writing, based on sound science, from a perspective upholding ecological reality—even when that conflicts with political reality. We provide a voice for the many effective but little-known regional and ad hoc wilderness groups and coalitions in North America and also serve as a networking tool for grassroots wilderness activists."

Wild Earth and the Cenozoic Society are the creation of Davis and Dave Foreman, cofounder of the environmental activist group, Earth First! "Cenozoic" refers to the latest era of geologic time, which is marked by the formation of modern continents and the diversification of birds, plants, and mammals. Davis and Foreman previously worked as editors on the *Earth First! Journal* but wanted to create an effort focused on conservation biology and proposals for large wilderness areas. *Wild Earth* has thousands of subscribers and also is carried in bookstores across the country. It is not laden with photos and advertising, as are many other environmental periodicals. Its roughly one hundred pages per issue focus on solid writing about biodiversity and wilderness and include book reviews, guest editorials, and letters to the editor.

The society's main focus centers on an effort called The Wildlands Project, which was discussed in detail in a special issue of *Wild Earth* published in 1992. Although other wilderness and environmental issues are discussed in *Wild Earth*, much of each issue is devoted to information about various efforts tied to The Wildlands Project. Operating under the theme of "rewilding the continent," the project's goal is to develop a proposal for a North American wilderness recovery strategy. "Existing protected wilderness areas are too small and fragmented to adequately protect biodiversity," Davis said. "The basic assumption behind the project is that all species have a right to their natural places in their ecosystems. The Wildlands Project is an attempt to provide habitat for all native species."

The project involves identifying large core wilderness areas—such as the Adirondack Wilderness in New York and the Greater Salmon Wilderness in Idaho—as well as habitat corridors connecting these areas. Cartography and text for the proposal have been developed region by region. The project's supporters believe modern society is approaching a crossroads for wildlife and wilderness as the turn of the century draws near. A comprehensive plan is needed to ensure the survival of America's wilderness in the future, Davis says. Although existing parks, wilderness areas, and wildlife refuges are a start, they are too isolated and represent too few ecosystem types to adequately protect all wildlife in North America.

Although opponents argue that current wilderness areas are sufficient, the number of grizzly bears in North America is dropping, and wolves had to be reintroduced to what was once their natural range in Wyoming and Idaho. With wilderness areas and connecting wildlife corridors set up in the Rocky Mountains, these animals could, theoretically, move freely from Canada to New Mexico. In addition, numerous other plants and animals face threats to their existence as more and more land is claimed for human development rather than remaining in natural states. For instance, only 10 percent of U.S. old-growth forest is still standing.

"The land has given much to us; now it is time to give something back— to begin to allow nature to come out of hiding and restore the links that will sustain both wilderness and the spirit of future human generations," the project's mission states. "The idea is simple. To stem the disappearance of wildlife and wilderness, we must allow the recovery of whole ecosystems and landscapes in every region of North America. Allowing these systems to recover requires a long-term master plan. A feature of this design is that it rests on the spirit of social responsibility that has built so many great institutions in the past. Jobs will be created, not lost; land will be given freely, not taken."

"This is a long-term project—meaning decades," Davis says. "There are many areas that are already wild and unsettled. We will identify additional areas that can be added to create core wild areas and corridors connecting them. As people leave an area or die, we see the land being protected by government or nonprofit groups. We are not proposing any forced relocation. This wildlands recovery strategy is an attempt to save all native wild species in North America. We think it's the most ambitious conservation effort under way on the continent."

A Wyoming nonprofit group's efforts, which have dovetailed with the Cenozoic Society's Wildlands Project, also center on love and respect for the land encompassing its bioregion in the southern portion of the state. The organization, Biodiversity Associates, and its ongoing struggle to preserve wilderness fit well with the goals of The Wildlands Project.

In 1988, while hiking in Wyoming's Medicine Bow National Forest, Leila Stanfield and a group of citizens from the Wyoming city of Laramie came across clear-cut after clear-cut. "It looked like a war zone," Stanfield said. "We got upset and decided to try and figure out why it was happening. We set out to determine what the Forest Service did and how the system of timber sales worked." To learn more, the group went on field trips with the U.S. Forest Service, the U.S. Fish and Wildlife Service, and Bureau of Land Management. They discovered there were avenues available for citizens to question and even halt timber sales that they considered unnecessary or destructive to the forest.

This learning process led to the formation of a group initially called Friends of the Bow, which focused on preserving the natural character of the Medicine Bow forest. Later, in a brouhaha that landed on the pages of a number of the state's newspapers, several representatives of Wyoming industry groups involved in livestock, farming, trucking, and mining incorporated the name "Friends of the Bow," in essence taking it away from its originators. Today, while still using the name Friends of the Bow on an informal basis, the Laramie-based environmental association operates as a nonprofit group incorporated as Biodiversity Associates.

Initially, the group started requesting maps and information about proposed timber sales from the U.S. Forest Service to learn more about the government's system for managing logging efforts in the nearby national forest. In addition, Stanfield, a private pilot, began flying over the forest to monitor logging operations. "By 1990, we had a shocking array of photos and discovered through Forest Service documents that the Medicine Bow was the most heavily logged and road-ridden forest in the southern Rocky Mountains," she says. "We started doing slide shows to raise awareness about the extent of clear-cutting and its impacts."

The aerial photography effort also revealed six main roadless and noncut areas remaining in the forest. The group set out to save these remaining parcels of forest even though the Forest Service's plan at the time slated logging operations on all six. By getting area communities involved, the group persuaded the Forest Service to institute a logging ban in the six roadless areas. However, the ban may remain in place only until a new forest plan is developed later this decade by the Forest Service. "Our concern is that the Forest Service only looks at the number of trees cut, not the fragmentation of the interior of the forest caused by clear-cutting—fragmentation that is harmful to a number of species," Stanfield says. "The Forest Service is supposed to protect these species. We want proof that the remaining pieces of forest are sufficient to ensure these species are safe."

Fragmentation of the forest disrupts natural corridors used by wildlife for migration, and clear-cutting simply eliminates large pieces of their habitat. In addition, clear-cutting can cause erosion problems when mountainsides are stripped of vegetation needed to hold soil in place, and the practice is not sustainable. Selective logging that removes only a percentage of the trees is at least somewhat sustainable because it leaves a semblance of the wildlife habitat in place and curbs erosion problems. Nonetheless, even with selective logging, roads must be built, which can lead to future problems in the forest by encouraging erosion and attracting additional logging efforts and more human activity.

While working to preserve the Medicine Bow's six remaining roadless areas, Biodiversity Associates came in contact with The Wildlands Project. The leaders of the large-scale wilderness project had heard of the Wyoming group's efforts at photographing and mapping and knew the group's knowledge of Wyoming forests could play a role in the evolving Wildlands proposal. In the early 1990s, the two groups joined forces to pore over maps and develop ideas on potential Wyoming core wilderness areas as well as corridors connecting them. Eventually, Biodiversity Associates presented written information to The Wildlands Project and continues to assist in the effort by attending periodic workshops and providing input on the intensive wilderness-vision maps being developed by The Wildlands Project.

Biodiversity Associates is continually struggling to fund itself through grants and private donations. It received a pleasant funding surprise in 1994 when the outdoor clothing manufacturer Patagonia turned a boycott aimed against it by a prologging group into a fund-raiser for the Wyoming environmental effort. The controversy swirled around a book published that year, *Clearcut: The Tragedy of Industrial Forestry*. The coffee-table-format book, published by Sierra Club Books and Earth Island Press, features more than one hundred full-page photographs of clear-cuts in North America that reveal the destructive nature of industrial forestry practices and essays about both the problems with and alternatives to clear-cutting. One two-page photo of Wyoming clear-cuts in the book was taken by Stanfield as she flew over the Medicine Bow National Forest.

The prologging group, called the Wyoming Resource Providers, took exception to the book. It urged people to call Patagonia on the company's toll-free order line and complain about the company's support and sales of the book. The group asked its supporters to end their calls by proclaiming they were boycotting the company, Stanfield says. However, when Patagonia heard of the boycott, it decided to donate ten dollars to Biodiversity Associates for every boycott phone call it received. Only 134 calls were made, many of them hang-ups, but the effort still tallied $1,340 for Stanfield's group's continuing efforts to preserve national forests.

Today, Biodiversity Associates immerses itself in a great deal of technical work as it works with the Forest Service on master plans for national forests in Colorado, Wyoming, and South Dakota, to be created over the next few years. "We're now trying to be more proactive and work with the Forest Service while these plans are in the development phase," Stanfield adds. "We are jumping in at the beginning of the process to try and effect change."

Naturalist Aldo Leopold made a point worth repeating. Too much of our land today is harmed because we treat it as simply a commodity to be controlled. More thought and planning needs to be given to the long-range future of our lands and their integrity. One way to accomplish this is by seeing the land as part of a community we belong to, as Leopold suggested. By developing love and respect for our lands, we will appreciate them for the benefits they provide our world, as well as the plant and animal worlds, in their natural states.

FOR MORE INFORMATION

Biodiversity Associates/Friends of the Bow, P.O. Box 6032, Laramie, WY 82070; (307) 742-7978

Planet Drum Foundation, P.O. Box 31251, San Francisco, CA 94131; (415) 285-6556

Wild Earth, P.O. Box 455, Richmond, VT 05477; (802) 434-4077

The depletion of the ozone layer and effects of global warming are no laughing matters. Thankfully, many scientists and companies are confronting these atmospheric hazards in creative and effective ways. (TOLES © 1994 The Buffalo News. Reprinted with permission of Universal Press Syndicate. All rights reserved.)

CHAPTER 16

FIGHTING FOR THE ATMOSPHERE

Most, if not all, environmental problems are tied to human practices, and issues involving the earth's atmosphere are no exception. Global warming—also known as the greenhouse effect—and depletion of the ozone layer are both complex and divisive environmental controversies.

Many scientists believe the greenhouse effect is caused predominantly by modern society's dependence on fossil fuels and wood. Many of the emissions caused by burning these fuels have become known as greenhouse gases, such as carbon dioxide and methane. They form a layer high in the atmosphere that acts as insulation and traps the earth's heat below it instead of allowing some of the heated air to cool at higher levels in the atmosphere. Although some greenhouse gases are needed in the atmosphere to help earth ward off the extreme cold of space, too large a greenhouse effect means the atmosphere retains too much heat, thus the danger of global warming.

Forests and plant life also play a role in this phenomenon. They serve as nature's way of absorbing carbon dioxide, but the earth's forests are now a fraction of their former size due to excessive logging and the need to harvest wood in many parts of the world for heating and cooking. Because of the universal burning of fossil fuels and the destruction of forests, many scientists believe an increase of greenhouse gases is occurring in the atmosphere, and studies on the dangers of global warming are gaining acceptance. A warmer world would have devastating effects, scientists claim, such as extreme weather patterns, diminished agricultural output, damage to and loss of numerous ecosystems, and the destruction of some communities by rising seas.

In September 1995, the 2,500 scientists who serve on the Intergovernmental Panel on Climate Control (IPCC) issued a warning concerning global warming. Since it came into existence in 1988, the IPCC had never stated its concerns so plainly and unanimously. The scientists agreed that the planet's climate was becoming unstable, the effects of which are likely to cause widespread economic, social and environmental dislocation over the next century.

The depletion of the ozone layer is a separate problem. The ozone layer sits fifteen miles above earth, is about twenty-five miles thick, and absorbs solar ultraviolet (UV) radiation. Many scientists fear a depleted ozone layer will lead to higher incidence of skin cancer and crop failure around the world. The ozone layer is in danger due to the use of harmful chemicals such as chlorofluorocarbons (CFCs), which are compounds that were once used widely as aerosol propellants and refrigerants. CFCs have been commonly used in automobile air conditioners and refrigerators, among other products.

In 1974, Dr. Sherwood Rowland and Dr. Mario Molina, chemists at the University of California at Irvine, discovered the ozone-depleting effects of CFCs, although their findings weren't readily accepted by governments or industry. Nonetheless, additional studies commenced to further the understanding of the earth's ozone layer, and the United States became the first country to impose limits on CFC use. In 1978, after consumer boycotts reduced the demand for spray cans in America by about 60 percent, the U.S. government banned the use of CFCs as propellants in most aerosol spray cans. In 1995, Rowland and Molina, along with Dr. Paul Crutzen, professor at the Max-Planck Institute for Chemistry in Mainz, Germany, were awarded the Nobel Prize in chemistry for their ozone-layer research work.

Unfortunately, the production and use of CFCs continues around the world. Third World countries still rely heavily on CFCs, and even the state of Arizona has argued for changes in U.S. regulations to again allow the use of CFCs in that state. However, evidence is mounting that damage to the ozone layer will lead to trouble. According to a U.S. government report dating back to 1975, ozone depletion of 50 percent would cause skin to blister after one hour of exposure to the sun. At this level, it also would be difficult to grow many food crops, if any. Since a hole in the ozone layer opened up over Antarctica during the 1980s, scientists have studied the effects of increased UV radiation on phytoplankton—the microorganisms that make up a vital first link in the food chain and maintain animal life in ocean waters. Preliminary indications show that phytoplankton populations under the ozone hole have dropped by nearly 12 percent. In addition, excess radiation streaming through a depleted ozone layer is thought to disrupt photosynthesis in most plants and cause cataracts in humans. Preliminary reports also point to increased incidences of blindness in horses, cattle, rabbits, and sheep in southern Chile, where high UV radiation already has resulted from the ozone hole over the South Pole.

In 1995, the World Meteorological Association reported that the hole in the ozone over Antarctica had spread to a size of 3.9 million square miles—roughly the size of Europe. Furthermore, the United Nations weather agency has been monitoring ozone levels in Antarctica for the last forty years. In 1995, the region had 30 to 35 percent less ozone than in 1950, according to the agency, and ozone levels over the United States and Europe had diminished by 10 to 15 percent since 1957. During the same time period, UV radiation had increased 13 to 15 percent.

Skeptics and scientists who disagree with the notion of ozone depletion claim that these findings are preliminary. They argue that the results could be related to a temporary drop in the amount of atmospheric ozone or are simply not accurate. With strong arguments being made on either side of the issues of global warming and the depletion of the ozone layer, solutions are proving to be very difficult to reach. This chapter discusses a few of the efforts two organizations are undertaking to fight for the conservation of the ozone layer and the limiting of greenhouse gas production.

In 1988, a local group in Olympia, Washington, formed under the name The Atmosphere Alliance to educate people about the greenhouse effect and the depletion of the ozone layer. After several years of operation at the local level, the group decided to shoot for a bigger impact by creating a National Day of Action for the atmosphere. The event was set for July 1, 1992, the first day it became illegal to vent CFCs out of refrigeration and air-conditioning equipment and into the atmosphere. "We ended up connecting with activists in forty cities and helped spark a number of demonstrations," says Rhys Roth, codirector of The Atmosphere Alliance. "And we ended up with a broader goal. Our mission now is to enhance the public's knowledge of our atmospheric crisis and mobilize grassroots action to protect the global climate and ozone layer."

Through various educational efforts, the group is trying to explain the science behind the atmosphere using the latest information and research. The alliance's long-term goal is to reduce global carbon dioxide emissions by 20 percent by the year 2005. One of its strategies is to supply people with the information they need to demand change. For instance, one of the alliance's tactics is to rebut "aggressive and well-funded misinformation campaigns" orchestrated by organizations financed by polluting corporations and industries by contacting the media and educating its members about these campaigns.

The group also is working to strengthen international atmosphere-related treaties such as the Montreal Protocol for Ozone, passed in 1987, and the Framework Convention on Climate Change, passed in 1992. The Montreal Protocol on

Substances that Deplete the Ozone Layer mandates the phaseout of CFCs by the year 2000 or sooner. It was signed by the twenty-four countries that produce and consume the most CFCs. Developing nations were given an additional ten years to meet the deadline because of the cost of conversion. If the agreement is followed, the ozone layer is expected to continue thinning until 1998 and then start recovering slowly and reach normal levels around the year 2045.

Concerning global warming, the United Nations Framework Convention on Climate Change was signed at the 1992 Earth Summit in Rio de Janeiro by the United States and other industrialized countries. The framework commits the countries to reducing emissions of greenhouse gases to 1990 levels by the year 2000.

At the local level, the alliance helps people manage three basic activities that can reduce energy consumption and air pollution. They involve building operations, land use, and material use. The group has several methods to help individuals and businesses save energy and water, as well as to reduce their waste production. To that end, it promotes programs where local power companies and social service agencies send energy auditors to analyze homes and buildings in an effort to make them more energy-efficient. Substantial energy savings have been realized with minor investments, and in many cases those investments are covered by government or other forms of funding. Less energy used means less fuel burned to create that energy, which means fewer greenhouse gas emissions, Roth says. Although his main goal is to protect the earth's atmosphere, Roth realized that to accomplish this goal he needed to back up several steps, examine how energy was consumed, and devise ways people could cut their energy use— and therefore emissions.

"There are two barriers to more efficient buildings—a lack of knowledge and a lack of financing. This effort addresses both of those problems," he says. "It also creates jobs, improves buildings, and improves the quality of life. One building in Ontario, for instance, has made a $26 million investment in hopes of saving $500 million in eventual energy costs. Of course, these energy savings also reduce greenhouse gas emissions."

The member roster of the nonprofit alliance has grown to more than five hundred. Memberships cost twenty dollars and include a subscription to the group's quarterly publication, *No Sweat News*. The publication is full of information for grassroots activists about how individuals can take action to aid the atmosphere. The group also has published a book, *Life Support! A Citizen's Guide to Solving the Atmospheric Crisis*. The book's goal is to bring the science of atmospheric issues down to an easily understandable level while also offering solutions to the mounting problems.

"By increasing people's understanding of the fact that we're in essence conducting this giant unknown experiment on the global atmosphere, we hope to make them aware of how critical the climate and ozone layer are to our survival," Roth adds. "And by giving them tangible actions they can take to help solve the problems and information about the situation, we hope this combination results in solutions. Ignorance is not bliss, and what we don't know can hurt us when it comes to altering the planet's complex life-support systems. The full consequences of ozone depletion and global warming are unknown, and we are performing inherently unsafe experiments that no sane scientist would ever propose to undertake. No sane society that cares about its children and the world should either."

As far as corporate efforts to protect the atmosphere are concerned, a new superefficient CFC-free refrigerator built by the Whirlpool Corporation won a "Golden Carrot Award" in an industry competition. The idea for the contest began in 1989 when representatives from utility companies, the Natural Resources Defense Council (NRDC), and the Environmental Protection Agency met to explore ways to reduce energy consumption. At the time, they knew refrigerator manufacturers were redesigning their products to meet approaching EPA guidelines banning CFCs and requiring more efficiency.

Since utility companies are finding it cheaper to promote energy savings rather than build new power plants, and because they know refrigerators consume as much as 20 percent of a home's electricity, they wanted to see refrigerator makers go beyond the new EPA requirements. Before long, the Super Efficient Refrigerator Program (SERP) came to life in the form of a nonprofit corporation funded by a group of twenty-four public and private utility companies nationwide. The goal was to bring more environmentally friendly refrigerator-freezers to consumers ahead of the time they normally would reach market.

In winning the $30 million prize in the SERP competition, Whirlpool had to compete with several other appliance manufacturers that had also developed super refrigerators. Whirlpool's winning design uses 25 to 50 percent less energy than 1993 federal standards require—a savings of between five and six hundred dollars over the typical twenty-year life of a refrigerator, says T. R. Reed, manager of financial communications at Whirlpool.

Unlike previous models, the new refrigerator accomplishes this without the use of CFCs. Additionally, computer technology allows the appliance to defrost only when necessary. The first SERP model—a twenty-two-cubic-foot side-by-side refrigerator-freezer—hit the market in 1994. "The progression in environmental benefits in our appliances has been great during the last five years," Reed says. "Apart from the money savings, these features make no difference to consumers, but the energy-efficiency and CFC-free features make a big difference for the environment. And consumers don't have to give anything up."

Whirlpool's $30 million award will be used to cover extraordinary design, development, and marketing costs—as well as more expensive materials and parts—so consumers will find SERP refrigerator prices similar to those of comparable models. SERP will pay Whirlpool as the refrigerators are delivered to participating utilities' service territories. Whirlpool expects to manufacture roughly 250,000 SERP refrigerators in a variety of models by 1997, when the program ends. "The real winners in this competition are the American consumers who purchase SERP products," Reed says. "They will benefit from the advanced technology and know they've helped to protect the environment. Ultimately, it's their purchase decisions that will determine the success of the SERP program."

Although the new Whirlpool design doesn't incorporate CFCs, the technology requires further modification to completely eliminate all chemicals harmful to the atmosphere. The first refrigerator-freezers are available in the Whirlpool, KitchenAid, and Kenmore brands. Whirlpool also developed one of the first devices to help automotive and appliance repairmen capture CFCs when repairing older cooling units. The airtight bags are designed to hold the CFCs from several refrigerators or automobiles until they can be safely extracted for recycling.

As the examples discussed in this chapter demonstrate, action to reduce our impact on the atmosphere can be taken at every level of society—from individual and group activism to corporate innovation. As Rhys Roth says, "We're conducting a giant chemistry experiment with our atmosphere." With the consequences that could be in store for our planet, can we afford to take chances?

FOR MORE INFORMATION

The Atmosphere Alliance, 2103 Harrison NW, Suite 2615, Olympia, WA 98502; (360) 352-1763

Natural Resources Defense Council, 1350 New York Avenue, NW, Suite 300, Washington, DC 20005; (202) 783-7800

Whirlpool, Administrative Center, Benton Harbor, MI 49022; (616) 923-3231

Emissions from cars, buses, trucks, and other internal combustion engines account for much of the air pollution we see and breathe today. Fortunately, people like Dr. Donald Stedman, shown here with his FEAT device that measures automobile emissions, are meeting the air pollution challenge head on. (Photograph courtesy of University of Denver. Used with permission.)

CHAPTER 17

WHEELS OF
ENVIRONMENTAL FORTUNE

A major cause of air pollution around the world is the internal combustion engine found in cars, buses, boats, trucks, tractors, and motorcycles. The world's first motorized vehicle dates to 1769, when Nicholas Cugnot, a French artillery officer, developed a steam-powered gun carriage, a three-wheeled machine that moved at one mile per hour. In 1885, German engineer Karl Benz put a gasoline engine in a vehicle that looked like a motorized tricycle. Of course, Henry Ford followed up these efforts with his popular Model T in 1908. This was the first mass-produced automobile, and it transformed transportation in America. Until 1912, however, 40 percent of the vehicles in the United States were steam-powered, 38 percent used electric batteries, and only 22 percent used gasoline. A simple device turned the tide—the electric starter. The Cadillac company began making cars with starters in 1912, and the invention made gasoline the fuel of choice for the automobile.

Thanks to these developments, vehicle emissions are now responsible for nearly 40 percent of the air pollution in the United States. Though emissions standards have drastically reduced the amount of emissions per vehicle, the number of vehicles on U.S. roads has doubled since 1970. The Environmental Protection Agency claims that just 10 percent of the cars on the road today produce 50 to 60 percent of total carbon monoxide emissions and 40 to 50 percent of total hydrocarbon emissions. Until new technology and public transportation alternatives change these trends, our best hope at reducing air pollution caused by gasoline-powered vehicles is to reduce the number of miles we drive, improve miles-per-gallon ratings, and decrease emissions. Fortunately, inventors and entrepreneurs are working on efforts to address these issues.

To help auto manufacturers boost fuel efficiency and cities meet stricter pollution standards, a Massachusetts company has developed what it calls "lean-burn technology" and is ready to introduce the system to the automotive industry. By increasing the amount of air mixed with gasoline before combustion in an automotive engine, the technology allows for more miles to the gallon along with a huge reduction in hydrocarbon and carbon monoxide pollutants, says Michael Ward, founder of Combustion Electromagnetics Inc. (CEI), of Arlington, Massachusetts.

Today's automobile engines run on a mixture of roughly fourteen parts air to one part fuel. With the lean-burn technology, an ordinary engine can burn mixtures of up to twenty-four parts air to one part fuel with little loss in performance. And best of all, a typical engine can be modified to "burn lean" at a relatively low cost. "This system will help the environment by giving car manufacturers the capability to set up their engines to achieve fuel efficiency and lower pollution levels at a very reasonable price," Ward says.

Although the lean-burn concept has been around for years, Ward is the first to make it work on current-production engines. During a recent independent test, a 1986 Ford Escort, retrofitted with the CEI ignition system, met proposed U.S. emission standards for the year 2004. Standards for carbon monoxide, nitrogen oxide, and hydrocarbons were all exceeded by a significant margin, and fuel consumption was cut by more than 10 percent.

Ward, who holds a doctoral degree in applied physics from Harvard University, became interested in lean-burn technology during the 1973 oil embargo. The peculiar characteristics of hydrocarbon flames piqued his interest to try to create an electrically stimulated low-polluting flame. "I ended up developing an approach that gives you the best of all worlds," he says. "By running an engine with a diluted air-to-fuel mixture, you can get higher performance for both emissions and fuel efficiency."

As a result of recent findings, which include test results on the company's technology by General Motors research labs in 1994 and additional testing conducted in Japan in 1995, older engines may benefit the most. While the lean-burn technology could be a boon to car manufacturers, it also could be used to clean up cars in high-pollution areas such as Los Angeles. Older cars in these areas could have their overall emissions reduced by as much as 90 percent at a reasonable cost, Ward says. CEI is now looking to form joint ventures with large automotive manufacturers and automotive electronics suppliers to commercialize the technology. The company already has been contacted by most of the large automakers and hopes to soon identify at least one to work with in the future.

Another auto-based pollution-reducing effort helps state officials get the worst polluters off the road. Amazingly, half of the air pollution created by vehicles today is produced by a mere 10 percent of all the cars and trucks on the road. These vehicles—typically older models or those whose emission-control devices have been altered—are the ones that get on Donald Stedman's nerves. To help combat air pollution and catch these offending vehicle owners, the chemistry professor at the University of Denver in Colorado has developed a smog-detection device that peers into traffic and finds violators.

The device, called a Fuel Efficiency Automobile Test (FEAT), is set up beside a roadway, where it can measure emissions from individual vehicles in less than a second. By shooting an infrared beam across the traffic flow at roughly a foot off the ground, the device measures the amount of pollution being released from a vehicle's exhaust pipe. A detector on the opposite side of the road measures how much of the infrared light travels through the exhaust stream. To catch offenders, a video camera completes the process by recording each vehicle's license plate number. The FEAT device can test up to one thousand vehicles an hour at speeds ranging from 3 to 150 miles an hour.

Stedman takes issue with annual emissions tests conducted by many states, where vehicles are tested while idling at certified service stations or while rolling on treadmills. A scheduled emission test for gross polluters is as effective as a scheduled breathalyzer test for drunk drivers, he argues. "Cars don't run at idle, and an annual test just makes people cheat once a year," he says. "I'd rather catch the offenders that are producing 50 percent of the pollution and take care of the problem." Gas taxes also aren't a solution, he adds. "If a person's job requires driving, the price of gas isn't going to matter. What will slip is the amount of money spent on maintenance. Gasoline costs more than four dollars per gallon in the United Kingdom, but cars there are still poorly tuned."

In a study conducted in Provo, Utah, Stedman's device detected polluting vehicles that were then repaired for free. The study revealed that the amount of gas saved due to the repairs could pay for the price of those repairs in only two years. The University of Denver holds several patents on the FEAT device, which is now capable of measuring carbon monoxide, carbon dioxide, hydrocarbons, nitrous oxide, speed, acceleration, smoke opacity, and a vehicle's operating temperature. The devices already are being used on Arizona highways, where government agencies issue tickets to the owners of offending vehicles. FEAT units also have been sold to the governments of England, Sweden, Thailand, Korea, Canada, and the United States, Stedman says.

In addition, the state of California also has leased a number of the FEAT devices and because of them has tested more than two million vehicles. In

August and September 1994, ten remote sensors made two million measurements for about fifty cents per car. In the process, half of the vehicles in Sacramento County were analyzed—many of them more than once. The effort revealed that 20 percent of the on-road pollution was from noncounty cars. This led the state legislature to authorize the use of FEAT devices to impose fines on violators. The devices cost about $100,000 each, comparable in price to a car treadmill, but are capable of measuring one hundred times as many cars.

"With better maintenance, you can dramatically drop emissions," Stedman says. "All these pollution-control measures based on the premise that all cars are equal don't work. I've created a way to help save fuel and clean up the air. You could identify all the dirty cars and fix them for free with the money now being spent on these other noneffective programs."

A completely different approach to fighting air pollution caused by vehicles lowers emissions while saving gas. Makers of a new engine treatment for automobiles claim their product can significantly reduce harmful emissions, cut fuel consumption, and extend the life of a vehicle's engine. Called Tribotech, the new product works simply by reducing the amount of friction occurring inside the engine. *Tribo* is the Greek word for friction—and friction consumes a significant amount of the energy produced by engines, says Nick Penta, chief executive officer of Simon Petrochemicals, a Simi Valley, California, company that produces the product. Tribotech was created by Juanito Simon, a chemist, metallurgist, and mechanical engineer who spent more than six years creating the formula. The product entails a patented mixture of castor oil, highly refined mineral oils, and ten other ingredients combined under high temperatures and pressure.

Numerous tests, both independent and company-conducted, have shown that engines treated with the product last longer, their hydrocarbon emissions are cut by an average of 75 percent, and their carbon monoxide emissions are cut by an average of 25 percent. Additionally, fuel economy increases by an average of 17 percent. "The product basically makes engines more efficient by reducing friction and heat, and in doing so cuts pollution and improves gas mileage," Penta says. "Tribotech creates a protective shield and cushion in an engine to make metal surfaces smoother and more wear-resistant."

Other engine treatments suspend solid particles such as Teflon in an attempt to lessen friction in engines, says Chris Lane, the company's marketing director. However, microscopic pieces of these solid materials can bond together and eventually end up damaging engines, he explains. One fourteen-ounce treatment of Tribotech will protect an engine for up to sixty thousand miles. A treatment retails for thirty-five dollars, but that amount can be equaled in fuel savings in just three or four months, Lane adds. The product currently is sold in NAPA and TRAK auto-parts stores in California, Arizona, Nevada, and the Midwest but should be available nationwide soon.

One California school district tried the product in several of its buses and soon ordered Tribotech for the entire fleet after seeing an amazing 55 percent reduction in diesel fuel consumption. "This dramatic reduction in the use of fuel will produce tremendous cost savings," wrote Bob Beal, the district's lead mechanic, in a letter to the company. "We are thrilled with the performance of your product."

In another test, Jerry Williamson, who owns an independent consulting firm in Cypress, California, had all the oil drained from a vehicle treated with Tribotech and a dry oil filter installed—then took off for San Francisco. "It was a real-world test. We drove 750 miles at seventy miles per hour in eighty-five-degree weather. And the car performed absolutely flawlessly. The engine didn't even heat up," Williamson says. "We would stop for gas and have attendants check the oil. They would come to us with these blank stares, and we'd say, 'Thanks, we'll have it looked at when we get back to L.A.'"

Of course, these are just a few of the ways of trying to make vehicles pollute less. Electric- and natural-gas-powered vehicles also offer promise and continue to grow in popularity. Drivers, too, can make a difference by keeping tires properly inflated and engines tuned. One of the best ways to prevent automobile pollution, however, is simply to avoid driving altogether. Walking, biking, carpooling, using public transportation, and relying more on the telephone can make a significant impact on protecting the air we breathe.

FOR MORE INFORMATION

Combustion Electronics Incorporated, 32 Prentiss Road, Arlington, MA 02174; (617) 641-0520

Simon Petrochemicals, 21 West Easy Street, Suite 108, Simi Valley, CA 93065; (805) 579-6608

University of Denver, Department of Chemistry, 2050 East Iliff, Denver, CO 80208; (303) 871-2580

American Forests volunteers admire the one millionth Global Releaf tree—a native acacia koa that they've just planted in Hawaii's Hakalau Forest National Wildlife Refuge. (Photograph courtesy of American Forests. Used with permission.)

CHAPTER 18

GOOD TREE-KEEPING EFFORTS

Trees are among the world's most precious resources. Forests provide critical habitat for thousands of plant and animal species and consume massive amounts of carbon dioxide—a greenhouse gas generated in part by human civilization that is believed to contribute to global warming. Forests also produce oxygen to help sustain life on the planet. Unfortunately, many of the forests of Europe and North America have been destroyed over the past few centuries. Now, developing countries around the world are cutting *their* forests for the sake of economic development and survival. Half of the world's population still depends on wood for basic needs such as cooking, light, and heat. Additionally, just one-quarter of the world's population consumes three-quarters of all processed paper and board—one of the leading causes of deforestation. This chapter highlights a few of the positive efforts currently working to keep the world and its trees alive.

The logging of ancient forests doesn't always receive as much publicity as do the threatened tropical rain forests of the world. But these old-growth forests located in the temperate regions of the earth—in places such as Canada, Siberia, and parts of the United States and Europe—also need protection, and a California nonprofit group is heading up the fight to save this critical habitat. Called Ancient Forests International (AFI), the nonprofit group is documenting the distribution of these vanishing ecological treasures and promoting their preservation, says AFI member Kathy Glass.

The grassroots organization began after its director, Rick Klein, worked as a park ranger in Chile. He kept hearing rumors of great redwoodlike trees, called alerce cypress, that grew in the country's southern regions. However, all he could ever find were stumps of the trees. After returning to the United States, Klein met loggers who had cut down alerce trees in Chile while working for U.S. logging companies. United States and Japanese logging concerns have operations around the world, harvesting large trees like the alerce wherever they can gain permission to cut them down. Klein decided to organize a group of people interested in finding some of the trees still standing. In 1989 the group traveled to Chile and hiked high into numerous valleys in search of the alerce trees. These rare trees can live more than three thousand years, a life span second only to California's bristlecone pines. Eventually the group found several large but isolated tracts of the huge trees in some of Chile's cool and damp ancient forests.

Since then, AFI has helped raise the issues of forest use and preservation in Chile and placed the legendary forests on that country's national political agenda, Glass says. The group also helped form Chile's first nongovernmental organization, *Fundacion Lahren*, which focuses its efforts on native forest issues. Through this Chilean group, AFI has helped direct fund-raising efforts to purchase more than one million acres of ancient forest in Chile. AFI itself purchased land for Chile's first private park—a 1,200-acre piece of forest in the country's famed Lake District. Called the Cani Forest Sanctuary, the park is dedicated primarily to environmental education efforts.

In addition, AFI is working on projects involving temperate forests in New Zealand, Tasmania, the Pacific Northwest, Poland, and Russia. In fact, Russia contains 20 percent of the world's forests. In Siberia alone, 30 percent of the world's remaining old-growth forests are found. However, this area is under increasing pressure by international companies like South Korea's Hyundai Corporation, which is clear-cutting on the Syetlaya Peninsula and threatening the existence of the endangered Siberian tiger. Old-growth forests are targeted by Hyundai and other companies for newspapers and toilet rolls in the United States, Europe, and Japan, Glass says. Clear-cutting is the practice of harvesting all the medium to large trees on a tract of land and usually killing all of the smaller trees and seedlings in the process. This logging practice also destroys the homes of most of the forest's creatures. The forest is difficult to regenerate because of the massive destruction left by clear-cutting and the ensuing soil erosion of the barren land.

"Temperate forests epitomize the cool, dark, cathedrallike essence of old-growth ecosystems, and the race is on to save the last of these ancient forests," Glass says. "They're all being cut, and anything we can do to spread awareness and create action is important. We feel very passionate about the issue."

In early 1995, the organization celebrated a victory. It helped introduce Doug Tompkins, an American environmental philanthropist, to a remote fjord in southern Chile called Cahuelmo. Tompkins set up a local foundation to hold title to the land and then purchased the fifty thousand–acre site to protect it from development. AFI also runs an active organizational and educational

center in the heart of redwood country in northern California. It works to conduct its own innovative environmental education activities and networks to help
other organizations in various efforts to save old-growth forests.

Another California-based forestry group, the Institute for Sustainable
Forestry (ISF), has been working to create a "good forest-keeping seal" through a
system aimed at certifying lumber from forests in the western U.S. that is harvested in an ecologically sound manner. The labeling program calls for landowners and logging operations to follow ten elements of sustainability in
harvesting their forests. To receive a seal of approval, the affected forests cannot
be clear-cut, doused with harmful chemicals, or torn up by an abundance of logging roads. Protecting biodiversity and enhancing local employment opportunities also are important criteria. "Realistically, we know we can't stop logging,"
said Jude Wait, executive director of the nonprofit effort. "What we need is a
more ecologically sound and sustainable way to do it. ISF is bridging the gap
between environmental concerns and economic realities."

The ten elements of sustainable forestry include the following practices
and efforts:

- Forest practices will maintain and/or restore the aesthetics, vitality, structure, and functioning of the natural processes of the forest ecosystem and
its components.
- Forest practices will maintain and/or restore surface and groundwater
quality and quantity, including aquatic and riparian habitat.
- Forest practices will maintain and/or restore natural processes of soil fertility, productivity, and stability.
- Forest practices will maintain and/or restore a natural balance and diversity of native species of the area, including flora, fauna, fungi, and
microbes for purposes of the long-term health of ecosystems.
- Forest practices will encourage a natural regeneration of the native plant
species to protect valuable natural gene pools.
- Forest practices will not include the use of artificial chemical fertilizers or
synthetic chemical pesticides.
- Forest practitioners will address the need for local employment and
community stability and shall respect workers' rights, including occupa-
tional safety, fair compensation, and the right of workers to collectively
bargain.
- Sites of archaeological, cultural, and historic significance will be protected and will receive special consideration.

- Forest practices executed under a certified Forest Management Plan will be of the appropriate size, scale, time frame, and technology for the parcel of land and adopt an appropriate monitoring program, not only in order to avoid negative cumulative impacts, but also to promote beneficial cumulative effects on the forest.
- Ancient forests will be subject to a moratorium on commercial logging during which time the Institute will participate in research on the ramifications of management in these areas.

The ISF labeling program, called Pacific Certified Ecological Forest Products (PCEFP), first requires landowners or logging operators to develop a timber-management plan. This plan provides a tree inventory, lays out long-term goals for the land, and describes how the ten elements of sustainability will be met. When harvesting is started, periodic inspections are undertaken by the institute along with the normal inspections conducted by state government. If all conditions are met, the lumber eventually produced will carry the PCEFP label, Wait says. By purchasing the certified and labeled wood, consumers know their buying power is supporting sustainable forestry and allowing them to influence forest-management policies. Lumber producers, in turn, have a marketing advantage through the creation of a market niche much like that enjoyed by organic food producers. The institute's efforts are being supported by forest advocacy groups along with the forestry establishment, including the U.S. Forest Service and the California Department of Forestry. "People who never used to talk to each other are now sitting down and agreeing on some plans and ideas," Wait says. "This effort shows we can start working together to get things done."

The idea for sustainable logging grew out of a company called Wild Iris Forestry in Redway, California. Owners Peggy and the late Jan Iris selectively harvested hardwoods on their land—as opposed to clear-cutting—and sold the kiln-dried wood for flooring and cabinets. The institute is taking the forestry system developed at Wild Iris and passing it on. The Holistic Forestry Service of Redway is one of a number of companies working with the institute. "You can't have ecological stability without economic stability," Wait says. "So in a lot of ways, this is a community-development project as well as an environmental effort."

In a separate effort, a unique tree-planting program would like to put a piece of America's history in your yard. The Famous & Historic Tree Program is an environmental education concept combining contemporary conservation with America's heritage. Young trees that are direct descendants of trees planted by or associated with George Washington, Betsy Ross, Martin Luther King, and 130 other famous people and places are available for planting. "We have identified trees all across America and around the world that are associated with significant people or events in history," says Neil Sampson, senior vice president of American Forests, the nonprofit group sponsoring the program. "From the seeds of those one-of-a-kind trees, we grow small healthy trees and make them available for sale."

An estimated 100 million tree-planting spaces are available around homes and businesses in U.S. towns and cities. Strategically planting those trees around buildings could save as much as $4 billion each year in energy costs by blocking the sun to reduce air-conditioning costs and winter winds to reduce heat loss. Those energy savings would reduce carbon dioxide emissions from energy production by an estimated 18 million tons per year, Sampson said, along with millions of tons of other energy-related pollutants such as sulfur dioxide and nitrogen oxide.

Included in the group's catalog are descendants of trees that witnessed the landing of Columbus, the American Revolution, and the bloody battles of the Civil War. Others were nurtured by presidents, inventors, artists, heroes, and other accomplished Americans. George Washington, for instance, planted numerous trees at his home in Mount Vernon, Virginia. The program's George Washington tulip poplar dates to 1785 and is the largest of the living trees planted by the first president.

Other famous trees are related to the lives of Abraham Lincoln, Ulysses S. Grant, Robert E. Lee, John James Audubon, Edgar Allan Poe, Helen Keller, Jesse Owens, Thomas Edison, and Henry Ford. Some of the effort's most popular selections come from Walden Woods in Concord, Massachusetts. Because this is where Henry David Thoreau lived and wrote from 1845 to 1847, it is considered a sacred spot of land by many. Singer Don Henley and other celebrities have helped raise money and awareness for American Forest's Walden Woods Tree Project, which is aimed at preventing development on the land around Walden Pond.

During 1995, the United States held many events to remember Franklin Delano Roosevelt, and American Forests marked the fiftieth anniversary of his death by offering two trees in his honor: a white oak from Hyde Park, New York, and a redbud, or southern magnolia, from Warm Springs, Georgia. "As governor of New York during the Great Depression, Roosevelt arranged for thousands of unemployed people to work on reforestation projects," Sampson said. "Therefore, these trees seem an appropriate symbol for FDR." The saplings offered by the program were descendants of Roosevelt's trees.

Founded in 1875, the American Forests organization itself is part of America's history. It is the country's oldest nonprofit citizens conservation organization. In addition to the Famous & Historic Tree Program, the group manages several diverse programs that revolve around trees, including another tree-planting program called Global ReLeaf, the Historic Tree Registry, which tracks the oldest living examples of numerous tree species, and an awards program to recognize extraordinary tree-planting efforts.

American Forests will host its seventh American Forest Congress in 1996 in Washington, D.C. The rarely held meeting traditionally has drawn government and industry leaders together with forest managers and scientists to develop a set of principles and recommendations for U.S. forests. The first such meeting was held in 1882 and the second was hosted by Theodore Roosevelt in 1905, and it led to the establishment of the national forest system.

For the sake of the myriad species that rely on forest ecosystems of all types for their very existence—as well as for the sake of our atmosphere—trees must be preserved and replanted around the world. Unfortunately, it becomes a complicated issue when jobs and the survival of families worldwide depend on practices that destroy forests. Nonetheless, modern society must find ways to preserve as many of the earth's remaining forests as possible.

You can make a difference by joining one or more of the many environmental groups that are working to save forests or conducting tree-planting efforts. In addition, by recycling waste paper and buying recycled paper you are saving trees. Planting trees in strategic places around your home is another way to help. By saving this most precious resource, modern society can dodge the potentially fatal accumulation of carbon dioxide in the atmosphere and preserve plenty of oxygen-manufacturing trees and forests for the future of our world.

FOR MORE INFORMATION

American Forests, P.O. Box 2000, Washington, DC 20013; (800) 320-TREE

Ancient Forests International, P.O. Box 1850, Redway, CA 95560; (707) 923-3015

The Institute for Sustainable Forestry, P.O. Box 1580, Redway, CA 95560; (707) 923-4719

ALTERNATIVE ENERGY

Wind Power Coming of Age

The Rising Sun

No Place Like Home

These spinning rotors are creating energy with wind—one of the cheapest and cleanest forms of renewable energy. (Photograph by Warren Gretz, courtesy of National Renewable Energy Laboratory. Used with permission.)

CHAPTER 19

WIND POWER COMING OF AGE

It drives golfers mad, puts millions of kites in the air, and has billowed a multitude of flags for centuries. It's a force so powerful that it can uproot trees and destroy small buildings. It's the wind.

It also has served as a power source for humanity for thousands of years. More than five thousand years ago, boats fitted with sails were driven by the wind on Egypt's Nile River. Later the windmill was invented, and the power it created was harnessed for numerous tasks, such as grinding grain and pumping water. By the 1300s, windmills dotted the landscapes of Europe. They were used in the Netherlands to help drain water off the country's low-lying land, as well as to serve various forms of commerce, such as the papermaking and wood-cutting industries.

The idea of using the wind to create electricity came much later, of course. In the late 1880s, a pair of men from Massachusetts were the first to set up small-scale wind generating stations in America. In 1890, a wealthy resident of Cleveland, Ohio, constructed a windmill that powered the more than three hundred lights in his mansion. Overseas, the Danish government began experimenting with electric production using the wind at about the same time.

And so, over the years, as the cost to produce wind-powered electricity has continued to drop, interest in these natural currents as an energy source has whipped up. Wind power has gradually grown in acceptance to become one of the most popular forms of alternative energy. Alternative energy refers to power derived from renewable sources—as opposed to the nonrenewable energy sources modern society uses today, such as coal, oil, and natural gas. Renewable sources of energy mainly involve the sun and the wind, although geothermal power is also being studied.

Although wind and solar power are typically considered two separate forms of alternative energy, in essence, energy derived from the wind is simply another form of solar power. The winds that sweep across the earth are created by the sun. It heats the surface of the planet but does so unevenly. Open areas become hotter than regions covered by forests or other vegetation, which remain cool. These areas of high and low temperature cause air to move back and forth, creating wind. Other factors, such as the presence of mountains and other specific land forms, also lead to more local wind patterns.

The true challenge for wind power is to be competitive with the costs of electricity produced by burning oil, coal, or natural gas. Like many other forms of technology, learning to do it better and more inexpensively takes time. However, although the costs of environmentally beneficial wind power have been dropping gradually for twenty years, they have been cut in half since 1990 due to technological advances.

Thanks to continued research and development, the price for wind power now hovers around five cents per kilowatt-hour at good sites. Including a federal cent-and-a-half tax credit currently allowed for wind production, the renewable power already is competitive with other conventional sources of energy. These new cost figures—along with a better grasp of the huge U.S. wind resource located mainly on the Great Plains, where high average wind speeds show exciting potential—have set the stage for a major commercial thrust in both utility-company and smaller wind systems, says Robert Thresher, director of the National Wind Technology Center, situated at the foot of the Rocky Mountains near Denver, Colorado.

"Wind is expected to be one of the least expensive forms of new electric generation in the next century. At good sites, wind power will be competitive without tax credits by the year 2000," he says. "We're pretty close, but the industry needs continued support in research and development efforts because most wind companies are small businesses that are just starting out. They don't have the strong support base and cash flow of fuels like coal and oil. Right now, many of these guys are having a tough time doing the research to get to the next generation while at the same time installing current technology and trying to be competitive."

Private companies in California currently produce the lion's share of U.S. wind power—more than 1,600 megawatts of wind electricity. Another 500 to 1,000 megawatts are under discussion in the state. (One megawatt of electricity typically will power a community of one thousand people under average weather conditions.) Across the United States, more than fifteen thousand wind turbines are producing a total of about two thousand megawatts of power.

Most of these turbines, installed more than a decade ago, do not benefit from newer technology that has boosted the power produced per turbine by roughly ten times. However, utilities and wind developers have announced plans to build wind sites to produce more than 4,200 megawatts of new wind generation by 2006 in fifteen U.S. states. Developers around the world put more than twelve hundred megawatts on line in 1995 alone. As for the future, in blustery North Dakota alone estimates say wind sites could produce as much as 30 percent or more of the amount of electricity the United States consumed in 1990.

To a certain extent, wind turbines rely on technology similar to that used in standard power plants. Coal, for instance, is burned to create steam in many of these plants, which in turn spins a rotor. The spinning rotor then creates electricity through an electric generator. Instead of steam, wind turbines use the force of the wind to spin a smaller rotor and create electricity.

New blade designs are continuing to increase wind turbines' potential energy capture. Many new turbines now rely on computer-controlled systems that vary the angle at which the blades meet the air. Called variable-speed turbines, these models can keep their rotors spinning at optimal speeds in a wide variety of wind conditions. Frequent and severe changes in wind speeds hampered earlier constant-speed turbines, resulting in the need for more expensive and heavier components, as well as higher maintenance costs.

To continue bolstering wind-turbine research and development, the National Wind Technology Center opened its doors in 1994. "The center is attracting scientists and manufacturers who share the dream of widespread economical wind power," Thresher says. "Ultimately, we hope companies will seek to develop the vast wind resources of the Great Plains."

Run by the U.S. government's National Renewable Energy Lab (NREL), the center will accommodate up to sixteen wind turbines and allow numerous companies to research and test their various wind products. Four turbines already have been constructed at the center, and when it is fully operational within the next five years, up to five megawatts of power could be generated. A wind-technology facility at the site was originally built in 1981, but when the Reagan administration slashed the renewable-energy budget in the mid-1980s—believing the research was a waste of money—the facility ended up in the hands of the U.S. Department of Energy's (DOE) weapons program. However, the Clinton administration provided a resurgence in the renewables budget, so the site was refurbished and reoccupied in 1994. The center encompasses nearly three hundred acres of land at Rocky Flats, a DOE facility near Denver that once produced plutonium-bomb triggers during the Cold War.

NREL liked the site because it sits in an area that experiences two distinct wind patterns, Thresher explains. During the fall and winter, powerful winds sweep over the Continental Divide and down the Front Range of the Rocky Mountains, reaching speeds of more than one hundred miles per hour and producing severe conditions in which to test the durability of wind turbines. In the spring, consistent and smoother winds blow in from the east and northeast, providing operating conditions similar to those on the Great Plains, where the bulk of the nation's wind resources are located. Scientists and engineers can use computer models at the center to simulate operating wind turbines and individual components. The ten thousand square feet of laboratory space at the facility is large enough to allow for the disassembly of large turbines so that their individual components can be analyzed and modified.

"The site is designed not only to be a center for research, but a technology magnet for a new industry as well," Thresher says. "It is a place where NREL scientists are working side by side with wind-turbine developers to create the advanced wind-turbine systems of the future—and where wind-plant operators and utilities can come for technical assistance. Our goal is to perform research on wind technology and assist the industry in making wind power a cost-effective option for utilities as well as at remote locations."

Although wind power's fuel is free and it does not create the numerous environmental problems associated with many other forms of energy production, it does kill birds that fly into the spinning blades of wind turbines. In many cases, wind farms and raptors live in close proximity because both operate best in open spaces.

A two-year study called for by the California Energy Commission and completed in 1992 considered the impact of wind turbines on birds at California's Altamont Pass, the site of the world's largest wind farm. Here, thousands of fifty-foot-high turbines occupy twenty thousand acres of land fifty miles east of San Francisco. Of the 183 dead birds found there, 111 were raptors. These birds of prey included red-tailed hawks, American kestrels, and golden eagles. Just over 50 percent of the deaths were attributed to collisions with turbines, but the extrapolations made in the study produced some potentially damaging news for the wind industry: Too many golden eagles could be dying because of it. Besides being less abundant than other raptor species in the area, golden eagle breeding rates also are inherently slower, meaning turbine deaths could be trouble for the species. However, because bird deaths are hard to record and track, no one is quite sure of the true impact wind farms are having on raptor populations. The wind industry states that according to its research about one raptor a year is killed for every one hundred turbines nationwide.

Statistics like these put the U.S. Fish and Wildlife Service in a tough position, says Mike Jennings, a biologist working for the service in Wyoming, where a new wind farm is being considered on the state's windy southern plains. "It's confounding because we support wind power but are responsible for enforcing laws set up to protect various bird species. You support the idea but must suggest modifications that slow its progress. It can be a dilemma."

To shed more light on this quandary, another study is under way to look at the effects of wind turbines on the overall golden eagle population around Altamont Pass rather than just on the number of deaths, says Thresher. This study will determine estimates of the total eagle population in the area as well as of transient eagles that travel through the region sporadically. It then will estimate the yearly mortality rates caused by turbines and the resulting impact on the overall eagle population. The National Wind Technology Center and the Audubon Society are involved in the study, which is being conducted by the University of California at Santa Cruz.

More than one hundred golden eagles of all ages have been tagged with radio-transmitters and are being monitored. When one of the birds' transmitters does not move for more than four hours, a mortality sensor is activated, warning of a possible death. From March 1994 through July 1995, eight deaths had occurred—only two caused by wind turbines. "The only way to really get at what's happening to the birds is to look at their overall population—is it growing or getting smaller?" Thresher says. "The population is the main thing. Wind farms may end up occupying land that otherwise would be developed and end up being more harmful to the species."

Nonetheless, steps already are being taken by the wind-power industry to minimize the impact wind farms have on local bird populations, particularly various species of raptors. In particular, California-based Kenetech Windpower—the largest developer of wind-energy systems in the world—is involved in an ongoing program to reduce impact on birds. The company, which is facing financial difficulties, has spent more than $2 million on research and specific improvements to make its wind farms safer for birds of prey. For example, the company plans to use long, straight tubular towers for its turbines instead of the lattice-type towers installed in the past. The lattice structures attract the birds because they provide numerous perching spots. Perch guards are now being placed on these older structures to prevent the birds from landing on them.

In addition, Kenetech has avoided putting ladders and catwalks on the new tubular towers, which had provided perching spots in the past, and instead will conduct maintenance on the turbines from bucket trucks similar to those used by telephone and other utility companies. Other innovative techniques include painting the turbine blades in patterns or different colors to make them more visible to raptors and other birds. "Patterns will be installed on the blades used at the Wyoming site," Ken Jennings says of the project, which remains in limbo because of Kenetech's financial problems and the bird mortality issue. "When spinning, we hope the blades will look like a solid visual obstruction to the birds. We know we'll never completely eliminate bird mortality, but we're working at minimizing wind power's impact. We hope they cut mortalities and keep them so low that it's not a problem. We're working with industry to get the best technology in place with each new facility."

Although much of the United States' wind power currently is produced in California, the wind-power industry as a whole is primed to expand, says Alexander Ellis, Kenetech's vice president of marketing. Many other areas of the country offer equal or better wind sites than does California, and the DOE estimates that wind energy could eventually supply nearly 20 percent of the United States' electricity needs. "We are certainly keeping our eyes open to all parts of the country in trying to expand wind-driven power plants," Ellis says. "We've identified thirteen states that have even better wind than California."

To help bring more wind farms into existence, Kenetech has designed and developed a concept it calls a "Windplant," a large array of interconnected wind turbines that operate as a single power plant. This computer-based system permits automatic and independent operation of each single turbine while also controlling hundreds or thousands of wind turbines as a unit. With this system, plant capacity in the form of turbines can be added in increments as energy demand grows, and individual turbines can be serviced at scheduled intervals without necessitating that the entire plant be shut down.

Developing nations also stand to benefit from wind power. A single stand-alone wind-power system can provide enough energy to significantly raise the standard of living for many of the Third World's poorest rural people, Thresher says. One turbine with battery storage capability can handle services such as refrigeration, water pumping, water treatment, grain grinding, communications, and power for a school or community center. "We have a small but active program for Third World applications aimed at using small wind turbines," Thresher says. "We also are studying a hybrid system—a powerhouse that uses renewable energy from the wind and sun along with batteries and a diesel generator. This backup generator would run the system when sufficient wind or solar resources are not available."

Called a "village power system," the unit can be shipped overseas in a steel shipping container that is then simply loaded onto a truck for delivery. These hybrid systems cost less in the long run, are better for the environment, and provide improved reliability over stand-alone diesel generators. "These people have no power now, and renewables can provide that power," Thresher adds. "We're getting a lot of interest from developing countries. In essence, it involves bettering people's lives."

As more huge wind farms and smaller stand-alone systems are constructed in the next decade, providing clean energy to more and more people, the costs of producing wind power should continue to drop. And through the ongoing research and development driven by these new projects, the feasibility of wind-generated electricity will become more and more apparent. Wind power does not harm the environment, provided raptors manage to avoid the spinning turbine blades, and its fuel is free, clean, and renewable.

FOR MORE INFORMATION

Kenetech Windpower, 500 Sansome Street, San Francisco, CA 94111; (415) 398-3825

National Wind Technology Center, c/o National Renewable Energy Laboratory, 1617 Cole Boulevard, Golden, CO 80401; (303) 384-6900

Solar energy is free of harmful air pollutants that are produced by its fossil fuel counterparts, and it can be "harvested" a number of ways. Suntek's Cloud Gel, with its unique sensitivity to subtle changes in temperature and light, helps regulate buildings' internal environments effectively, affordably, and without creating pollution. (Photograph by Wendy Walsh, courtesy of Suntek, Inc. Used with permission.)

CHAPTER 20

THE RISING SUN

Although the sun is more than 93 million miles from earth, it provides another free and plentiful source of clean fuel for renewable energy production. Solar power does not generate the tons of air pollutants produced by fossil-fuel generated electricity. As a matter of fact, people have tapped into various forms of solar energy for thousands of years. The famous Greek mathematician Archimedes allegedly destroyed a Roman fleet in 212 B.C. by using a system of mirrors to focus the sun's rays on the ships and set them on fire. More than a thousand years later the ancient Anasazi Indians of the American Southwest built their cliff dwellings facing south in order to permit the sun to warm them during the cold winter months.

Heating modern homes with solar power is accomplished with one of two general systems—passive and active. Passive solar systems are simple and non-mechanical. In many cases, passive heating involves the use of a glassed-in space, such as a patio or porch, to collect sun-generated heat that is then dispersed to other parts of a home. Active solar systems for homes, on the other hand, often rely on mechanical solar panels. Inside a panel, sunlight heats a black surface material that in turn heats a liquid, such as water, in pipes. The heated liquid is then pumped inside the house to meet various needs, such as for space heating or hot water. Photovoltaic cells, which produce electricity directly, create another form of solar power.

One project in California is creating a huge new type of active solar system—one that could someday produce large amounts of electricity. The Electric Power Research Institute (EPRI) is working with Southern California Edison on the Solar Two project, which is developing methods of collecting heat energy from the sun for the production of electricity. The $40 million project under construction in the California desert uses thousands of computer-driven mirrors that track the sun and focus beams of sunlight onto a central receiving tower, explains Ed DeMeo, manager of EPRI's solar-power program. This receiver absorbs solar energy and heats molten salt that has been pumped to the top of the tower. Later, the super-heated salt is used to turn water into steam, which then spins a turbine generator to produce electricity.

In an earlier project in the 1980s, appropriately called Solar One, water was pumped to the top of a receiving tower, where it was turned into steam. However, in this system the steam had to be used immediately to produce electricity, and any cloud cover would slow the system considerably. So EPRI has turned to hot salt in its follow-up project. "The salt acts as a storage medium for the heat generated by the sun," DeMeo explains. "In this project, we use two tanks of hot salt. The cool tank is kept at four hundred degrees—well above the minimum temperature needed to keep salt in a liquid form. After the salt circulates through the receiver, it hits the hot tank at one thousand degrees." Because the physical properties of molten salt allow it to stay hot much longer than steam can, the improved system is much more efficient. And by separating solar collection from electric generation, clouds have less of an impact on the system. The superheated salt can be used over time to create steam for electricity production, which means several hours of cloud cover won't affect the process negatively.

Although electricity generated by Solar One and Solar Two is not yet cost-effective, DeMeo says the idea still holds promise: "Any time you can generate electricity from the sun, you're not generating it from a fossil fuel, which means no emissions." The only environmental drawback to the idea is the large amount of land needed to install the thousands of mirrors needed to capture the sun's energy. Continuing research and development work in the coming years will demonstrate whether this form of solar electric production can be cost-effective.

Another California-based utility has taken a different approach to solar power. Pacific Gas and Electric Company (PG&E) decided to use the sun to help solve some of its problems in meeting peak summer electricity demand. In a joint project with EPRI, the company built a huge photovoltaic system on five acres of land near the city of Fresno, according to Brian Farmer, PG&E's project manager. "At full power, the system produces a half of a megawatt of electricity," he says. "And the time frame works well for us because the system is generating power when our demand is peaking in the middle of the day."

The ability of sunlight to produce electricity—called the photovoltaic (PV) effect—was first observed in 1839 by Edmond Becquerel, a French physicist. He experimented with producing minimal amounts of voltage between two metal plates immersed in fluid and exposed to sunlight. His discovery essentially went unnoticed until the U.S. space program revived PV-related research efforts in an attempt to power spacecraft. In 1958, a U.S. satellite's operations were powered by a PV solar panel. As the space program expanded, PV solar cells proved to be more reliable and much lighter than other types of spacecraft power systems. PV cells typically are composed of silicon, one of the most abundant elements on earth. When sunlight strikes silicon atoms, its energy knocks electrons loose. In simplest terms, the movement of these electrons produces electricity. Simple devices such as pocket calculators now rely on PV technology for their limited energy needs.

For the PG&E project, eighty large solar panels have been set up at the project site, and inside each large unit are fifty smaller PV panels that automatically rotate to follow the sun's path and convert sunlight to electricity. This project marks the first time a PV system has been used to add more power to a utility's system, Farmer says, and not just to serve a certain customer or small group of customers. The project, which produces enough electricity for a community of five hundred, is aimed at determining whether PG&E can save money not by adding transformers and other electrical equipment to meet peak loads, but by relying instead on solar power. "If utilities can avoid system upgrades, a lot of costs can be avoided," Farmer says. "This effort will help us find out if that is possible. Even if solar power's operating costs are high now, it's the way of the future because costs should eventually come down."

Connected to PG&E's effort is Photovoltaics for Utility Scale Applications (PVUSA), a research-and-development organization working on expanding the use of solar power. To date, PVUSA has helped with seventeen PV projects in six states. And although these projects are intended primarily for research and demonstration purposes, they *are* supplying usable power to America's power grid, albeit in small amounts. So far, these projects have supplied enough power to satisfy the electricity needs of about nine hundred homes for one year. PVUSA's main goal is to determine the potential of PVs in large-scale applications. Because PV cells with twenty-five-year lifetimes are now manufactured—taking the place of earlier versions that had only five-year life spans—PVUSA work is proving to be cost-effective. Since the technology is silent, emission-free, and requires no fuel or cooling water, it involves negligible environmental impacts and regulatory risks.

On a smaller scale, a revolutionary "solar sphere" approach to PVs has the potential to produce low-cost clean energy for the homeowner. This idea involves imbedding tiny balls—or spheres—of silicon on aluminum foil to produce a thin, skinlike solar panel. This breakthrough in PV technology is the result of a ten-million-dollar, six-year effort between Southern California Edison Company and Texas Instruments (TI), the Dallas-based manufacturer of computer equipment and consumer electronics products. "The unique characteristics of this technology, coupled with its potential for low costs, offer a broad range of possibilities," says Jim Skelly, TI's solar program manager. "The material could literally become the skin of many buildings, being integrated into surfaces like the roof of a home rather than being an attachment, as is typically seen today."

By combining inexpensive and abundant materials with a low-cost manufacturing process, TI believes the technology can produce affordable solar systems for new homes and businesses. The target cost for a system using these new PV cells is between $1,500 and $3,000, far less than the $8,000 to $12,000 of today's technology. This is mainly because TI's new approach uses inexpensive low-purity silicon that costs about one dollar a pound instead of the thirty-five-dollar-a-pound semiconductor-grade silicon used in other PV-cell production. Southern California Edison and TI have developed a patented process using this low-cost material to create individual four-inch-square solar-sphere cells. Each square is composed of seventeen thousand tiny silicon spheres imbedded on a thin piece of aluminum foil that makes the finished product flexible, lightweight, and durable. This flexibility allows the cells to be molded into a variety of shapes not previously possible with large conventional cells. They are just as effective as other PV cells and don't require the extra protection of glass, saving more than 65 percent of the weight incurred on conventional modules, Skelly says.

The aluminum serves as both the panel's structure and as the conductor for the electricity produced by the spheres. And because each of the seventeen thousand spheres operates independently, the impact of individual cell failure is negligible when compared to today's cells. Once commercial manufacturing begins, the new solar cells could be installed on new homes and buildings. Systems for existing structures would be developed later as manufacturing costs drop. "One could imagine a number of potential applications, such as rooftop shingling, independently powered street signs, even self-charging laptop computers," Skelly says. "Lightweight flexible modules could easily be attached to virtually any kind of dwelling, making them particularly well suited for many remote locations."

TI originally began developing the spheral solar technology in 1983, drawing on its expertise in engineering and metallurgy. The company began producing its first solar modules for testing and evaluation in the early 1990s. The modules have been tested outdoors for several years with no measurable problems. "We believe this technology is ready for full-scale manufacturing," Skelly says. "It has been through development and pilot-line production. Now, a buyer in the energy market or a supporting industry can take the next step and move it into the marketplace." During the past several years, TI has focused its research on microelectronics and information technology. Because of this focus on its core business, the company has decided to sell its spheral solar technology to a company more closely involved in the energy industry.

Still another type of new solar technology responds to heat and light and regulates the amount of sun that enters a building. This "solar shutter" uses the sun for heating and lighting or repels it for cooling. The development could revolutionize the housing industry and drastically cut current levels of energy consumption. Suntek, Inc., based in Albuquerque, New Mexico, has created a substance it calls Cloud Gel. The gluelike mixture is placed between two layers of clear glass or plastic film that are then used for skylights, greenhouses, sun rooms, and nonviewing windows.

Cloud Gel remains perfectly clear when the building is cold or when light levels are low, thus allowing as much sun in as possible. However, it turns into an opaque, reflective white to block the sun when it's warm or extremely bright. This change from clear to opaque can occur with only a two-degree rise in temperature, says Suntek's president, Roy Chahroudi. "With this concept, we cut out many of the negatives of solar," he says. "It uses the very skin of a building—supplying both heat and light or rejecting excess heat and light—in a very simple process." By varying its ingredients, the gel can be produced to change opacity at any temperature between 68 to 104 degrees Fahrenheit.

Suntek has incorporated the use of Cloud Gel into a new product it calls the Weather Panel—a prefabricated building component. With a roof made of Weather Panels, a passive-solar house is simple to design and build, and the dwelling can look conventional and carry a conventional price tag. Along with Cloud Gel, Suntek's Weather Panel also uses several other solar technologies invented by the company to save energy. The lightweight panels are roughly three inches thick and are produced in sizes up to four feet wide and six feet long. They are entirely prefinished and begin functioning upon installation.

Suntek plans to sell Weather Panels to various building materials manufacturers, who will then incorporate them into their own product lines. Several companies are now testing the panels, Chahroudi says, and the technology should be licensed to building-material manufacturers in 1996. "We're a company of only ten people, but we've developed a product that could save one-sixth the world's energy consumption at market saturation," he says proudly. "We're in this to enable world energy consumption to be reduced."

These solar-power efforts are only four of a multitude of projects focusing on harnessing more of the tremendous energy constantly being created by the sun. Although tapping this source of power has proven difficult so far in terms of the costs involved, strides continue to be made. And since each and every day the sun showers the earth with more energy than is found in all the world's fossil fuel supplies, capturing some of it is definitely a goal worth pursuing. With enough research and effort, solar energy may some day in the future entirely replace the fossil fuels we rely on to provide power for modern society.

FOR MORE INFORMATION
Electric Power Research Institute, 3412 Hillview Avenue, Palo Alto, CA 94303; (415) 855-2159

Suntek, Inc., 5817A Academy Parkway East, Albuquerque, NM 87109; (505) 345-4115

Texas Instruments Solar Program, P.O. Box 655012, MS 35, Dallas, TX 75265; (214) 995-2011

CHAPTER 21

NO PLACE LIKE HOME

Alternative energy is playing an increasing role in the housing market as revolutionary energy solutions are incorporated into more and more construction efforts. These new design concepts are allowing the power of the sun to heat homes across the country, reducing their reliance on energy produced by pollution-causing fossil fuels. Along with helping the environment by relying on the sun instead of coal, gas, or oil, new home designs can save homeowners money in the long run.

The germination of Michael Reynolds's unique concept for a new type of house sprouted from a television program in the late 1960s. While watching Walter Cronkite's news broadcast, he saw a story about steel cans. At that time, beverage containers were made of steel, not aluminum, and the reporter mentioned how the cans—particularly beer cans—were being thrown everywhere and causing a trash problem. Cronkite followed this news story with a piece on the timber industry and housing. With all the new houses and buildings in construction across the country, America's forests are headed for trouble, predicted Cronkite in closing the broadcast.

"Those two stories sparked an idea in me," Reynolds remembers. "I was just out of the University of Cincinnati's architecture school and had done my thesis on housing. Within days of that broadcast, I was working on a building block made of steel beer cans." He succeeded in building a house of beer cans by incorporating a concrete mixture into the building process. But he then decided to try a new tack, using old tires filled with dirt instead of cans as the outer walls of his next house, which he dubbed an Earthship. The tires created three-foot-thick walls that provided superior insulating qualities to the thinner steel-can walls. Reynolds eventually developed a new type of home that does not rely on virgin materials to build or operate and is allowing homeowners the chance to construct and own a completely self-sufficient dwelling anywhere they please. He has since founded a company called Solar Survival Architecture in Taos, New Mexico, to promote the concept.

To complete the outer walls of an Earthship, old tires are filled with dirt and are then rammed by a sledgehammer until the dirt is tightly packed. Then, layers of the filled tires are piled up until a U-shaped frame is created. Each dirt-filled tire acts like a four-hundred-pound brick encased in rubber and provides excellent insulation. An Earthship needs no energy for heating or cooling because of these thick walls. Although it may take several months for an Earthship to reach 70 degrees inside initially, once it reaches equilibrium with the temperature of the earth it will maintain that temperature indefinitely. To give the house a more conventional look, the tire walls are covered with stucco or adobe mud.

Each home faces south, and a glass wall conducts the solar heat inside the dwelling. Interior walls for an Earthship use Reynolds's initial idea of steel cans with a concrete-mixture covering. "The use of recycled materials as building products is a direct benefit to the environment," he says. "And they make perfect building materials."

Hundreds of Earthships have been built across the United States as well as in Japan, Bolivia, Australia, and New Zealand. Hundreds more are now under construction. The homes are simple enough to construct that practically anyone can build one. A family of four in Minnesota once built an Earthship in two months for roughly twelve thousand dollars using an instruction book published by Solar Survival.

Mechanical systems within the homes also benefit the environment. Solar energy can be used for power and light in the dwelling, a rainwater-catchment system can provide water, and by using solar or composting toilets, the need to connect to a septic tank or sewage line is avoided. Also, by incorporating a greenhouse into the structure or simply using floor space on the south side for plants, Earthship owners can grow their own food. Reynolds suggests using a system to collect "gray water"—used water from sinks, showers, and washing machines—to nourish the fruit and vegetable plants.

"Over the years, as new environmental problems unveiled themselves—such as food, water, energy, and waste—I incorporated solutions to them in the Earthship design," Reynolds says. "The ultimate goal was to create completely self-sufficient homes. Earthships can now heat and cool themselves, make their own electricity from the sun, catch their own water from the sky, take care of their own sewage, and on top of that, grow food for their owners. So we're not taking anything out of the earth, and we're not putting anything bad back into it. We call them Earthships because you can basically build them anywhere, and they can be completely independent and take care of themselves. In the last few years, it has all come together."

Solar Survival Architecture offers a number of ways to help interested people learn more about Earthships. Periodic training seminars are conducted, and several books have been published to provide information about the structures and give practical advice on their construction. In addition, blueprints for various types of Earthships are available, as is assistance in constructing one of the homes.

In California, instead of filling tires with dirt, David Easton is using a simple mixture of soil and cement to create fashionable new homes that will last for centuries and cost little to heat or cool. By reviving an idea used by ancient Chinese and African builders, the renowned builder and Stanford University engineering school graduate has built more than a hundred of his "rammed-earth" homes. He expects growth in his company, the Terra Group, to boom in the coming years. "As environmental awareness continues to grow in our society, we see the future for earth-wall technology to be very, very bright," he says. "These homes are an environmentalist's dream come true."

Solid-earth construction spread in Europe courtesy of the Romans centuries ago. It became an important building technique in France, where the practice became known as *pise de terre*, which means "rammed earth." Easton borrowed the term and formed an acronym for his rammed-earth process: PISE, Pneumatically Impacted Stabilized Earth. The process creates walls that match the strength, texture, and color of sedimentary rock.

An Easton dream home is unique, to say the least. First, a typical concrete foundation is poured before reusable wooden forms are set in place on top of the foundation. Then begins the methodical construction of two-foot-thick walls made up of a mixture of 90 percent soil and 10 percent cement. For many homes, the soil that is excavated to accommodate the foundation can be used, or soil can be purchased inexpensively from a nearby quarry. Six inches of the mixture is placed in the wooden forms at a time. Using tampers—jackhammers equipped with a flat-foot attachment—the mixture is pounded until it is as dense as stone. The process is then repeated until the home's walls take on a marblelike quality similar to smooth sandstone. A four-person crew can complete up to 1,200 square feet of wall per day.

The eighteen- to twenty-four-inch-thick walls, which are naturally water-resistant and get stronger with age, can accommodate both electric and plumbing systems, as well as standard door and window units. No paint is required for the walls, which require little or no maintenance. When topped with a nonflammable roof a rammed-earth home is essentially fireproof—a big plus in many areas of California—and the thickness of the walls also provides exceptional protection from noise. Construction costs are only slightly higher than other construction methods, Easton says, but savings in utility bills soon make up the difference.

A rammed-earth house is much more environmentally sound and energy-efficient than is a conventional structure. The massive walls provide such effective cooling that no air-conditioning system is needed. And in California, where most of Easton's houses have been built, cooling costs represent 50 percent of a homeowner's utility bills. Additionally, a rammed-earth house uses half the amount of wood needed in conventional homes. And no plaster, fiberglass, or drywall is required to form the outer layer of the house.

Best of all, a rammed-earth house has the potential to last forever. "If you can picture all the millions of homes built every year—in fifty or sixty years most of them will finish their serviceable lives and have to be destroyed and buried in landfills," Easton explains. "With earth homes, there will be no burying or rebuilding. Because the soil is packed so densely, rain won't affect these homes. They will literally last for centuries. My goal is to get the earth-wall technology into the mainstream of the housing industry. We have a lot of misconceptions to deal with, but once people see what a solid building it is and feel the comfort and security the walls provide inside the house, they're sold."

Salt, of all things, supplied the impetus for another type of home that spares the environment and has won a U.S. Department of Energy award for innovation. It started one evening when Michael Sykes sat down on a pile of driftwood on a North Carolina beach. He was amazed at the heat the wood still retained well after dark. Sykes realized that the salt in the wood was helping hold in the heat. Being a builder of log homes, he thought he might have struck on an idea for a new way to heat homes.

Soon after this experience, he formed a company called Enertia Building Systems in Wake Forest, North Carolina. The company builds homes using salt-impregnated logs and a unique design that cuts energy costs for heating by 90 percent over conventional homes. "When I initially built log homes, I noticed that they retained heat for three or four days," Sykes explains. "That led to the idea of combining salt-treated wood and either solar heat or a wood stove to create a house with little or no heating requirements."

A prototype Enertia home in New Hampshire was compared to a conventional home next door. The Enertia home had no heating costs during the winter, but the house next door needed three thousand dollars in heating oil. The

More and more efficient, alternative-energy homes are built each year. The Aquarius II in Jackson, Wyoming, is a combination solar/geothermal home from Enertia Building Systems. (Photograph courtesy of Enertia Building Systems, Inc. Used with permission.)

property that allows mass—in this case from wood—to retain heat is known as "thermal inertia." Sykes decided to change the first letter in his company's name to "E" to symbolize the environment and energy.

An Enertia home relies on a unique air loop for heating, created by a double-shell design. Air is circulated through the outer shell of the house to distribute heat evenly. On the south side of an Enertia home, an eight-foot-wide room is built with glass on the outer wall and logs on the inner wall. Air is heated in the space between the two and rises, recharging the home's thermal battery. On the north side, the space between two log walls is only eight inches, and as the air in the space cools, it is pulled down and eventually circulated back to the south side for warming. A simple ceiling fan moves the air, and solar heat stored in the salt-laden logs is slowly released over time, keeping the loop in motion.

Enertia houses are made from twenty-year-old pines bought from nearby tree plantations, Sykes says, so no old-growth forest is destroyed. These logs also replace the need for the drywall, vinyl siding, and insulation used on the outer walls of conventional homes. So far, thirty Enertia houses have been built in more than a dozen states, and the company is building a factory to mass-produce the structures. The houses are precut, and the timbers are numbered grooved, and packaged in the order they will be used. Homeowners must put in a conventional basement foundation and install a roof of their choice.

"Our effort helps the environment because we drastically cut the use of petroleum-based housing materials in our construction," Sykes says. "And we're replacing houses that use plenty of energy for heating and cooling—pulling them off the market one by one."

Our homes, because of their design and construction and the fossil-fuel-based energy needed to heat and cool them, tax the environment, and there are millions of them on the planet—with more and more being built every day. By relying on new ways to build, heat, and cool them, the environment, as well as the people who dwell in them, will benefit.

FOR MORE INFORMATION

Earthships/Solar Survival Architecture, P.O. Box 1041, Taos, NM 87571; (505) 751-0462

Enertia Building Systems, 13312 Garffe Sherron Road, Wake Forest, NC 27587; (919) 556-0177

Rammed-Earth Homes/Terra Group, 1058 Second Avenue, Napa, CA 94558: (707) 224-2532

Natural Products

Environmentally friendly clothing is becoming more fashionable. Here, Sally Fox inspects the warm organic hues of the bolls of Fox Fibre's naturally colored cotton. (Photograph by Cary S. Wolinsky, courtesy of Natural Cotton Colours, Inc. Used with permission.)

CHAPTER 22

CLOTHING THAT
MAKES A DIFFERENCE

For a while there in the Garden of Eden, clothing wasn't an issue. Thanks to a snake and an apple, however, that situation changed soon enough. Today, human society makes all types of clothing, and what fashion designers determine people want typically takes precedence over the environment. Wastewater is created by the dyes, bleaches, and other chemicals needed to process cloth into garments. The agricultural and industrial efforts required to produce the raw materials for clothing add to the environmental burden through energy use, the production of waste materials, and the use of chemical pesticides and fertilizers.

However, there are alternatives in clothing production that benefit the environment, or at least cause less harm than traditional methods, by either cutting back on the amount of waste and pollution created in various manufacturing processes or by practicing recycling and turning what used to be a form of waste into a usable product.

In Arizona, an innovative entrepreneur is producing cotton that is beneficial to the environment because it grows in living color. What began as a plant-breeding hobby eventually turned into an environmentally beneficial business for Sally Fox. She developed plants that grow colored cotton right on the stalk and now sells her product to companies such as Esprit, L. L. Bean, and Levi Strauss & Co., sparing those firms the need to bleach and dye cotton before manufacturing clothing from it.

Fox first worked in entomology, researching ways to reduce the use of pesticides in farming. In 1982, while breeding cotton plants for pest-resistance, she came across some seeds covered with a beautiful brown lint. Excitedly, Fox asked the cotton breeder she worked for if she could research the possibility of growing colored cotton. He responded that no one wanted brown cotton and that she would have to do it on her own time. So she searched for seeds with nicely colored lint and planted them on a small plot of land. Although the plants produced colored cotton, as she had hoped, the fibers were too short for spinning machines used in the manufacturing of cloth.

A lengthy cross-pollination breeding program followed as Fox bred the colored plants to white plants with longer, higher-quality fibers. By 1988 she developed machine-spinnable colored cotton that met many industry standards. Fox's breeding efforts continue today on fifty acres of land near Wickenburg, Arizona, as she works to get more and more colorful cotton plants to "settle down genetically." Plants not settled down grow very differently from one another, and inconsistent colors and fibers are produced, she explains. The breeding process is arduous and time-consuming. Out of 250,000 plants grown each season, Fox collects seeds only from the best 3,000 plants for the next growing season.

"We now have literally thousands of lines of plants and have created a very rich and diverse program, but it takes at least seven generations of plants to gain enough genetic stability to produce high-quality fibers," she says. "Some plants need twelve or even thirteen generations." After each season, the fibers from the best plants are tested for length, strength, and other qualities important for spinning. After testing, seeds from just one-third of the best three thousand plants are planted the following season: 250 seeds from each of the superior plants. Fox made a huge breakthrough in 1995 when three new varieties of plants produced high yields of the most exceptional fibers to date. These new fibers will allow for faster spinning and overall processing.

After years of taking orders and not being able to fill them for a year and a half, Fox now has a warehouse full of bales of the older versions of her cotton. All future crops, however, will now bloom with the new varieties. Fox will take orders from various manufacturers wanting the newer cotton for their towels, linens, clothing, and upholstery, then hire farmers in Texas and Arizona to grow the crops. The cotton—sold under the trademark Fox Fibre—is available in varying shades of brown, green, and brownish red. Interestingly enough, the natural colors in Fox Fibre actually intensify after washing. The camel brown color *doubles* in intensity after twenty washes. This effect has to do with the natural waxes occurring in plants and trees, but Fox is still exploring the phenomenon.

To produce cotton cloth in the standard manufacturing process, white cotton must first be bleached to prepare it for dyes, Fox explains. After dyeing, finishing chemicals are used to set the colors so that they won't run. Dyes and chemicals cause two problems: They pollute water and cost money. In making yarn, if a textile mill pays a dollar to buy the cotton and another dollar to spin it, it will have to pay more than two dollars to dye it. Not only are the dyes themselves expensive, but plenty of energy, water, and wastewater treatment is needed to accommodate the dyeing process. In much of the Third World, where wastewater treatment is not practiced, the dyes and chemicals used in the making of clothing add to the pollution problem because they are simply dumped directly into rivers and other bodies of water without any treatment. Cotton grown in natural colors eliminates the wastes created by these processes and saves companies energy, water, and dye costs.

"You can now have color without using any dyes—that idea has spurred our efforts," Fox says. "This also has made people reconsider the use of chemicals and how they can improve the production process from the farm to the store. We're still at the beginning of a new field, and we're just starting to understand all there is to learn. But it's definitely a step in the right direction."

The only real drawback to Fox's effort is that she must use increasingly scarce water to grow her own crops in the arid climes of Arizona. She understands that irrigating is a tough environmental sell for many people. Unlike the South, however, insects that attack cotton plants are not as prevalent in Arizona, which means a reduced reliance on pesticides. In California, where Fox began her venture, she installed water-saving irrigation equipment that cut her water use in half. But due to the lengthy struggle she has undergone just to keep her company, Natural Cotton Colours, Inc., alive over the years, she has been unable to purchase similar equipment for her Arizona operation, although that remains her goal. "The challenge is to do it sensibly and be as careful and environmentally responsible as possible," she asserts.

To that end, Fox's operation strictly follows organic farming practices. These measures allow the company to eliminate the use of toxic chemicals and fertilizers. "If you can farm without them, why not?" she says. "Organic farming is all about farming communities and the health of people, water, and land."

In the future, Fox hopes her cotton will be seen for what it is—a less-expensive alternative. "We've made a lot of progress, but my goal is that my colors replace dyes. And economically, they should. It is simply a cheaper alternative for any product. For certain color ranges, Fox Fibre provides a cost-saving option to make products in a way that doesn't add to the chemical burden."

Another step in the process of turning cotton into various products causes another kind of problem for the environment. In the manufacturing of cotton clothing, millions of cotton fibers fly free during the spinning process and fall to the mill floor. Typically, these fibers—called mill trimmings—are swept up and carted off to landfills. However, a New Jersey company is now collecting the trimmings and using them to make a line of recycled-cotton socks, T-shirts, and baseball caps. Founded in 1992, Take The Lead . . . And Step Into A Better World, Inc. already has sold more than 250,000 pairs of the recycled-cotton socks, along with 100,000 pairs of another line of socks made of organically grown cotton.

Better World—the often used, shorter version of the company's lengthy name—is the country's only recycled-cotton sock maker in the mainstream market. It receives the mill trimmings from clothing manufacturers in North Carolina, South Carolina, and Tennessee, says Loraine Kulik, director of Better World's community partnerships. The company cleans the trimmings and then respins them with virgin cotton to gain the correct fiber length to make socks. The mixture is 75 percent recycled cotton to 25 percent virgin cotton. "In using recycled cotton, we're saving landfill space by reusing a good material instead of wasting it—and we're encouraging recycling efforts by reinforcing the notion that everything can be reused again for another useful purpose," she says. "It hasn't always been easy, though. We have had to spend a great deal of time encouraging, inspiring, and convincing manufacturers to use both organic cotton and recycled cotton to make Better World products."

Five years ago, the cofounders of Better World started a nonprofit organization called LEAD, an acronym for Leadership, Education, and Development. The program supports college students working together on environmental and social issues in their local neighborhoods. The students design and teach courses to younger students on these issues and receive academic credit for their efforts. "But eventually the cofounders wanted to also create a 'for-profit' business to try and close the gap between businesses and family and environmental issues," Kulik says. "One of the cofounders' fathers—who had spent fifteen years in the sock business—joined the effort. So socks it was."

To further address social and environmental issues, Better World donates 10 percent of its pretax profits to the Child Welfare League of America and other charitable organizations helping children and the environment. The Child Welfare League of America is a seventy-five-year-old nonprofit group that works with more than seven hundred child-protection agencies around the country. The league provides assistance to its members through various training and program ideas and also works on legislative issues involving family and child-protection issues.

Better World socks can be found in more than thirty of the nation's department store chains, including Nordstrom, Bloomingdales, Dayton's, Dillards, and Levi-Strauss. They also are on the shelves of more than three hundred college and university bookstores, as well as numerous environmental products shops. The company currently produces seventeen women's cotton sock styles and six men's styles, all priced between five and eight dollars.

Cotton isn't the only material being recycled to make an impact in the fashion world. As millions of tires and inner tubes continue to languish in dumps and landfills around the United States, one person is trying to make a dent in the problem. Mandana MacPherson, founder of Used Rubber USA in San Francisco, produces a line of handmade and fashionable products made from used truck and bus tire inner tubes. Since its inception eight years ago, the line has grown to include purses, bags, belts, wallets, briefcases, notebooks, and even eyeglass cases.

The idea struck MacPherson while she attended art school at Brown University in Rhode Island. Carrying around art supplies one day, she upended an ink bottle and ruined a leather bag. In searching for a replacement that would resist such messes, she found nothing. Eventually she came across an old inner tube and created a prototype bag to carry her art supplies. Her friends liked it, and she ended up making several more, selling them in a few stores in Boston and New York City. "This was before the environmental movement was in full gear," she says. "People liked the black color, the shapes and styles, and the unusual textures in the rubber."

Unlike bags and purses made of vinyl or plastic, Used Rubber USA's bags are worth the extra cost needed to clean the rubber and create the products, because they will last a long time. In fact, the products come with a lifetime guarantee and are among the few durable and waterproof alternatives to leather. "It's comparable in price, but lasts forever. Some people still have the perception of it as trash, but acceptance is growing," MacPherson says. "We sell a lot of items strictly on style. We don't consider them a novelty—they're functional and fashionable."

Because there is no system in place to handle the collection of used inner tubes, Used Rubber USA serves as both collector and reuser of the waste product. The company works with a number of tire shops in the San Francisco area to collect the used inner tubes and goes through more than two tons of used rubber a year. Each product varies slightly in markings and texture due to the different patterns fond on inner tubes. The products incorporate rustproof, all-aluminum rivets plated with bronze, brass, or silver. The rubber can be cleaned with soap and water, and contrary to what most people might think, does not smell like a tire shop.

Clothing and fashion play important roles in nearly every society on earth. Because of this, the impact that clothing can have on the environment is enormous. As new and more efficient ways to produce clothing and fashion accessories are created, the environment will benefit, and being fashion-conscious will no longer mean one has to be environmentally unconscious.

FOR MORE INFORMATION

Better World, 2010 Center Avenue, Suite 1, Fort Lee, NJ 07024; (800) 532-3411

Natural Cotton Colours/Fox Fibre, P.O. Box 66, Wickenburg, AZ 85358; (520) 684-7199

Used Rubber USA, 597 Haight Street, San Francisco, CA 94117; (415) 626-7855

CHAPTER 23

SERVING A JUST CUP OF JAVA

Coffee is the world's largest traded food commodity and the primary export crop of more than sixty nations. As such, it is steeped in environmental and human-rights issues across the globe—particularly in the Third World, where nearly all coffee crops are grown. Too often in getting this massive crop to market every year in countries as diverse as Kenya, Brazil, and Indonesia, the coffee industry simply ignores these issues. Considering that it takes the average annual yield of one coffee tree to make one pound of coffee beans, the huge task of producing and harvesting this crop for coffee drinkers around the world requires the toil of millions of peasant farmers and farmworkers. A good portion of these workers face difficult lives—living without clean water, proper health care, adequate education, or decent homes.

On large plantations, coffee workers are paid minuscule wages for their labor and are treated in many ways like slaves. For instance, in some cases, migrant workers in Central America are housed in animal stalls without electricity, sanitation facilities, or running water. Many of the inequities these workers face are based in the societal problems established by centuries-old colonial rule, as is briefly discussed in chapter 8, "Helping Cultures Survive." In Guatemala, a military-ruled government answers to a small segment of the society—one that holds a firm grip on the country's wealth and political power. An estimated 2 percent of the population owns nearly 80 percent of the arable land in Guatemala, and a similar percentage rules the country's industrial base. This leaves a huge and growing portion of the population with few employment alternatives.

Most small-time peasant coffee farmers, although working for themselves, struggle to survive. They must deal with a worldwide industry controlled by influential and powerful businessmen. Roughly 8 percent of the retail price of coffee ends up going to the small farmer; a major portion of the price goes to shipping, processing, roasting, and distribution. And although they receive a similar percentage to that of smaller farmers for their coffee, plantation owners have economies of scale working for them and because of their influence are also involved in the shipping and processing of the coffee.

In addition to these human-rights concerns, most coffee farming today—like so many other forms of agriculture—relies on the use of toxic chemical fertilizers and pesticides, which raise environmental concerns. Small farmers must apply these dangerous chemicals themselves. On large plantations, farmworkers typically are responsible for applying the chemicals without the benefit of protective equipment. When used improperly, these chemicals can lead to cancer, birth defects, and a host of other health problems. The costs to land and water resources, as well as to the plant and animal worlds, also are significant, especially in countries where chemical use is not regulated and where older, more dangerous products are used. Despite all of this bad news, however, several efforts are being made to improve these unfortunate conditions in the coffee-producing world.

On the nonprofit side of the issue, a group called Coffee Kids, formed in 1989, is helping change the coffee industry so that families and communities in the coffee-growing world are given the chance to improve their lives. Not surprisingly, children in coffee-growing communities are swept into the turmoil of their parents' struggle to survive. This stems from economic necessity, because many families must put their children to work as soon as farms are willing to pay them or they are able to help on the family farm. Children in the coffee-growing industry typically begin working by the age of ten, sacrificing their education from the third grade on.

In 1984, Coffee Kids' founder Bill Fishbein started a small coffee shop in Providence, Rhode Island. He finally started making money from the venture a few years later. Success, however, brought a feeling of uncertainty into his life, a feeling he couldn't quite pin down. He decided to take a vacation to Cape Cod and read several John Steinbeck novels, finishing with *The Grapes of Wrath*. "That book carved something into my heart. I saw a connection between the characters in the novel and the migrant coffee workers who were responsible for my financial independence," he explains. "I decided to do something."

During a subsequent trip to Guatemala, Fishbein witnessed the situation firsthand. What he discovered there appalled him and changed his life. Living and environmental conditions in the coffee-growing areas he visited were terrible. Water sources were polluted, health care was nonexistent, and education opportunities for children were extremely limited. He also was shocked at the difficulty faced by families just to survive. When he returned to the U.S., Fishbein started Coffee Kids, in 1989. Initially the organization worked with another nonprofit organization, Childreach, to get businesses and individuals to sponsor children whose families depended on coffee for a living. The money donated goes to food and other basic family needs. "I saw it as a way to give something back to these communities," Fishbein says. "It's a nice way for people to involve not only their business, but their employees and customers, too. Through letters, a dialogue and sense of community can be created between the children and their sponsors."

Although Coffee Kids still sponsors hundreds of children, its outreach has broadened. Now the group also raises funds through hundreds of coffeehouses, specialty roasters, wholesalers, and importers operating primarily in the United States and Canada. Money is collected from these businesses and their customers through Coffee Kids membership dues, general contributions, coin-drop donations, and the sale of promotional materials such as T-shirts and coffee mugs. These funds are applied to numerous projects in nine coffee-growing countries and involve issues such as sanitation, health care, nutrition, education, entrepreneurial efforts, and the environment. "Our work is to bridge the gap between consumers and producers," Fishbein says.

All these projects are founded on the principle of self-reliance and address the social, economic, and environmental problems faced by various coffee-growing communities. The responsibility for determining priorities and managing projects always remains with the community. Health-care programs promoted by Coffee Kids train local women in preventive care and focus on clean water, proper nutrition, and traditional herbal medicines. Education efforts help build schools and supply much-needed educational materials to the communities. In Mexico, several schools have been built using bricks made from coffee pulp—a by-product of the drying process that would otherwise be dumped as trash. In the Mexican town of Coatepeque, women mixed the pulp with clay to form bricks. The community then pitched in to build the school. Coffee Kids donated money to the effort to purchase doors, windows, and other hardware for the school building.

The creation of village banks is another idea supported by the organization. These banks provide credit to women to start small businesses such as chicken farms, fruit and vegetable stands, and weaving or other handcraft enterprises. The additional source of income helps families lessen their dependence on coffee as the sole source. More than thirty banks already have been funded.

For example, in the Guatemalan town of San Lucas Toliman, on the shores of Lake Atitlan, thirty-six women formed a village bank. Some used their loan capital to prepare and sell foods such as tortillas in the local market while others became dressmakers. The opportunity to develop leadership and business skills has given the women the means to achieve greater self-sufficiency, Fishbein says, and has allowed them to supplement their families' basic diet of corn and beans with other foods. The men tend to concentrate their efforts on coffee-growing.

To further help these small farmers, a new type of coffee dryer, which will provide more freedom in the marketplace, is being supported by Coffee Kids. Coffee-drying techniques have changed little over the past century, but this small solar-powered dryer allows small farmers to dry their freshly picked coffee berries within the twenty-four to thirty-six hours before the berries spoil. Once they are dried, the berries can be stored until the farmer chooses to sell them. The new dryer is being tested in Guatemala and Honduras. If it proves efficient and reliable, a loan system similar to the village-banking idea will allow small farmers to purchase or lease the dryers for their farms.

Finally, Coffee Kids works with local farmers to help them better understand the detrimental effects of erosion and chemical use, as well as the benefits of naturally increasing the health of the soil. For instance, alley cropping is a method of farming that cultivates various crops in the rows between coffee trees. Planting peas or beans in the rows is beneficial to the coffee trees, which are nourished by the nitrogen legumes return to the soil. The farmers and their families benefit, too, from a more varied diet.

Organic coffee-farming efforts are promoted by Coffee Kids. In fact, the communities the organization works with in Mexico now grow only organic coffee. "These people have always thought the only way to produce enough coffee was to use a lot of chemicals, but they're rethinking this practice," Fishbein says. "Through our efforts, we are trying to increase awareness in the coffee-consuming world that there is a real benefit to organic coffee—a benefit to the people who grow the coffee." Organically grown coffee tastes no different from other types of coffee, and through brochures distributed to the public and discussions with business members, Coffee Kids hopes to increase awareness about and sales of organic coffee.

"Helping communities in all these various efforts is a complicated and delicate business," Fishbein says. "It must be done with respect so dignity is not lost, dependency not created, and cultural aspects not disturbed." To that end, Coffee Kids makes sure each community it works with decides what projects it wants to tackle and how the efforts will be undertaken. Some communities—still reeling from some of the past effects of development projects that provided little benefit to the local people—are quite suspicious when first contacted by Coffee Kids. For instance, one Guatemalan community stayed aloof for several years after first being approached by the organization. "No thanks," they said. "We've seen this before." Much later, when the people realized the nonprofit group would not force its plans on the community, they agreed to work with Coffee Kids. "We want to promote the process of sound and respectful development, but we must go slowly and at the community's pace," Fishbein explains. "We're in it for the long run and have to have the patience to sow the seeds and gradually let them germinate."

Promoting sound development is also the goal of a U.S. company involved in the coffee business. The people who own and operate the Thanksgiving Coffee Company don't want you to just have a cup of coffee, they want you to enjoy a *just* cup, as in justice for coffee farmers and the environment. Through its coffee sales, the company supports organic-coffee growers in various parts of the Third World, including Asia, Africa, and Latin America. By working with collectives and cooperatives to ensure the coffee is grown without chemical pesticides or fertilizers, the Thanksgiving Coffee Company helps farmers use sustainable agricultural methods, such as using wasps to fight dreaded coffee-damaging beetles. "More than 40 million gallons of pesticides are used each year on coffee crops around the world, not to mention all the chemical fertilizers that are used," says the company's founder, Paul Katzeff. "Certified organic coffee is a way for me to take my ecological and social concerns, weave them into one package, and create a socially and environmentally sound approach to gourmet coffee."

Katzeff, who earned a master's degree in social work in 1970 but got started in the coffee industry soon after, decided to tie his social responsibility to the coffee business. As chairman of the Specialty Coffee Association of America in 1985, he developed a political and social theme for the group's national convention. Soon a special brand of coffee—Coffee for Peace—was created to raise money for the National Union of Ranchers and Growers of Nicaragua. "The coffee industry had tried to remain apolitical, but in fact it's all about politics because the business is tied to so many *campesinos* working with coffee in the Developing World," he says. "The coffee industry should be about helping people, not about living off the backs of people who work hard and get nothing for it."

The coffee business is involved in politics because if U.S. and other coffee buyers from around the world simply turn their backs on the plight of so many Third World *campesinos*—peasant farmers and farmworkers—they are, de facto, supporting the human-rights violations committed and supported by many governments and plantation owners. Injustice is injustice—whether it happens to your neighbor or to a peasant farmworker who is treated like a slave in Central America.

To help small-scale coffee farmers, the Thanksgiving Coffee Company pays between 40 to 75 percent above the market value for their organic coffee beans. Because no chemical fertilizers or pesticides are used, organically grown coffee requires more labor-intensive farming methods, including tasks such as composting and alley cropping. Of course, this does mean that the company must charge higher prices for its products.

By supporting small cooperatives, where much of the world's ecologically responsible coffee is produced, the company is also helping small farmers band together and compete with large coffee growers on an equal basis in the marketplace. Working together, small farmers can combine resources to handle their processing and transportation needs. In addition to the money it pays farmers for their crops, for every package of coffee sold the Thanksgiving Coffee Company returns fifteen cents to the growers through nonprofit organizations, such as Coffee Kids, working to improve working and living conditions in the coffee world.

In the United States and Canada, the Thanksgiving Coffee Company promotes the use of reusable plastic-mesh coffee filters to help reduce the amount of paper filters used in brewing coffee. More than 160 million pounds of paper coffee filters end up in U.S. landfills every year, Katzeff says. "I see my job as promoting change in the coffee industry. And because coffee is the second-largest traded commodity in the world [behind petroleum], there is a lot of potential for change."

On top of all these efforts, Katzeff helped found the Coalition for Sustainable Coffee. The group promotes a wide range of socially beneficial ideas and practices in coffee-growing regions, including many of the projects mentioned earlier in this chapter. One of its new projects involves developing a seal of approval for endorsing fairly traded, organically grown, "green" coffee. "Fairly traded" coffee is coffee grown by small farmers who work together in cooperatives. Although the standards for the seal of approval are still being developed, Katzeff himself applies thirteen different criteria to the so-called green coffee he buys. At the Thanksgiving Coffee Company, if a farm scores ten points from the criteria, its coffee is eligible for purchase. The largest point-scoring measure, which tallies five points, involves growing coffee the traditional way—in the shade, under the rain forest canopy. The obvious benefit to this growing method is that the native rain forest is preserved, meaning numerous species of plants and animals are protected. Scoring another three points is certified organically grown coffee, and sun-dried, small-farm-grown, and cooperative-member-produced coffee add one point each.

By promoting the use of a seal of approval, which could be based on Katzeff's criteria, the coalition hopes to prompt more U.S. coffee importers to operate in a socially responsible manner and help consumers understand the workings of the coffee industry in the Third World. "Since most coffee is exported to the industrialized countries of the north, retailers and consumers can play a large role in how coffee is marketed, and therefore, how coffee is produced," Katzeff explains. "By enabling consumers to choose sustainable coffee through their daily purchases, we can advance the consumption and production of fairly traded and environmentally friendly coffee."

If more coffee drinkers start demanding organically grown coffee and spend their money on coffee produced by small farmers and farm cooperatives, the tide may start to turn in favor of equity for peasant coffee farmers and farmworkers around the world. Although these "green" coffees are not prevalent in the marketplace now, if more U.S. coffee companies see that consumers will pay a little extra for coffees from, say, the Thanksgiving Coffee Company, they too will begin selling and promoting them. This could eventually force the large-scale plantations in the coffee-growing world to treat their workers more fairly in order to sell their coffee to discerning coffee buyers. Though this is obviously not an easy evolutionary process to undertake in a centuries-old industry, it is one well worth pursuing.

FOR MORE INFORMATION
Coffee Kids, 207 Wickenden Street, Providence RI 02903; (800) 334-9099

Thanksgiving Coffee Company, P.O. Box 1918, Fort Bragg, CA 95437; (800) 648-6491

The operations of any company can have a negative—or significantly positive—impact on the environment. Tom Chappell of Tom's of Maine is among the vanguard of environmentally and socially responsible American business leaders. (Photograph courtesy of Tom's of Maine. Used with permission.)

CHAPTER 24

ENVIRONMENTAL ETHICS
IN THE BUSINESS WORLD

Corporations around the world consume massive amounts of natural resources in producing products and offering services. The result is huge amounts of waste. In the way they run their operations, companies of all sizes and types can have a tremendous impact on the environment—for better or worse.

Sadly, too many corporations still believe that the consumption of natural resources and production of pollution is not a problem worth being overly concerned about. Yes, many more corporations today take matters of pollution into consideration, but others continue to take advantage of lackluster regulation and enforcement to secure greater profits at the expense of the environment. Industry worldwide pours a constant stream of various forms of pollution into our air, water, and land. Though this may be unavoidable to a certain extent, in too many cases rules are stretched and broken. And if companies *are* caught and hauled into court, too many times their legal teams manage to blunt the consequences of breaking pollution-control laws. In the case of conducting business in the Third World, too often, regulations simply don't exist.

But on the other hand, there is a growing theme that is attracting more and more businesses. It is based on the concept of social responsibility—a way of doing business that brings environmental concerns to the fore and also takes into account such issues as workers' rights and community service. Some companies are cutting their energy bills and waste-disposal needs as a way to save both money and the environment. Others are using less packaging in their products to reduce waste and costs or are relying on recycled materials for their products and packaging. These socially responsible companies conduct business with ethics in mind, and in doing so make money while making a difference for the environment.

Since its beginning, Tom's of Maine has included environmentalism as a company cornerstone. In 1968, founders Tom and Kate Chappell left Philadelphia to move to rural Maine as part of their desire to get back to the land. Although they ate natural foods and used natural products whenever possible, the Chappells were unable to find any natural personal care products. The idea for a company soon germinated. Their first product, called Clearlake, became the nation's first nonphosphate liquid laundry detergent. At the time, phosphates in wastewater were inducing algae growth that choked a number of American rivers. The Chappells peddled their product in the early mornings to dairy farmers as well as to the rest of their area's residents. They encouraged customers to recycle their empty containers by mailing them back—postage-paid—to be refilled.

Soon the Chappells had created a line of personal care products, and Tom's of Maine was born. In 1975, Tom's Natural Spearmint Toothpaste paved the way to a prosperous future. The company now produces natural soap, toothpaste, deodorant, shampoo, mouthwash, shaving cream, and dental floss. Natural ingredients play an important role at Tom's as the company continues to replace artificial ingredients. Saccharin, for instance, is an artificial sweetener found in all major brands of toothpaste—except Tom's, where cinnamon, spearmint, and peppermint oils are used instead.

From its modest door-to-door beginnings, Tom's of Maine has grown to become a business that generates $17 million a year in revenues. Although the company is still privately held, its products are now sold in seven thousand health-food stores nationwide, along with twenty thousand supermarkets and drugstores, says Tom Chappell, president and CEO. The company even exports its products to Canada, Israel, Saudi Arabia, and the United Kingdom.

Over the years, care for the environment has continued to play an important role in all Tom's of Maine's products. The company's toothpaste tubes are recyclable because they're made of aluminum. And all packaging is made from 100 percent recycled materials. "We eliminated the outer packaging for our mouthwash products and just attached an accordion booklet to the container to convey our message. This reduced packaging by 90 percent, which reduced our costs," he says. "In turn, we were able to lower the cost to our customers—which helped sales."

The company now offers the first refillable deodorant roll-on container. The new refill package provides enough deodorant to fill a permanent applicator bottle twice, resulting in a 20 percent reduction in solid waste each time a refill is purchased. Tom's Natural Spearmint Toothpaste remains the company's best-selling product, but a children's version named Silly Strawberry is gaining popularity and may soon become number one. The company makes another unique product for children—a bar of soap made to fit their smaller hands.

"We care about the environment, and we live by that commitment," Chappell says. "We have a strong focus on respect for nature and people. It starts with our corporate mission, which drives the company's philosophy and strategies. It's about trying to do the right thing—being financially responsible while being environmentally sensitive and socially responsible."

As if running a groundbreaking corporation wasn't enough, Chappell went back to college in 1986 to earn a master's degree in theology from Harvard Divinity School. "I did this for two reasons," he explains. "First, I wanted to understand my role in life. Secondly, I wanted to see how a business might otherwise be imagined, based on religious philosophies. After four years of theology school, I had the ammunition to make sure we were running an ethical enterprise." His studies clarified for him the need for more socially responsible practices and efforts in the business world.

Chappell's efforts have gained the attention and respect of others. In June 1995 he won the Socially Responsible New England Entrepreneur of the Year Award, sponsored by *Inc.* magazine. "Other nominees came from companies that were bigger and growing faster," he says. "We won because we're the values company—it shows that the way you do business is important." Chappell is confident that his ethical approach to business is rubbing off on other companies, too, which may encourage them to adopt similar practices.

Additionally, Tom's of Maine dedicates 10 percent of its profits to charity and other causes. Three projects funded by the company in Brazil are designed to encourage preservation of the rain forest through research and development of medicinal rain forest plants. By developing these kinds of sustainable harvesting practices, economic alternatives can be created for native people while saving the rain forest.

Tom's of Maine products do not contain animal ingredients and are tested for safety without the use of animals. Tom's requires a written guarantee from its suppliers stating that none of the ingredients supplied to the company are tested using animals. "If a company truly cares about the environment, it has to go the extra mile," Chappell asserts. "As Tom's of Maine gets bigger, we will make a bigger impact on society." As it grows, the company will carry more clout in terms of the volume of its natural products on store shelves, the number of charitable efforts it can support, and its overall influence on the business world.

A similar corporate case study is found in California. The Patagonia Company is dedicated to making high-quality outdoor clothing and equipment but is also devoted to maintaining the quality of the outdoors. "We're committed to minimizing our environmental impact," says Patagonia's Lu Setnicka. "We're also trying to maximize awareness about the critical point that the natural world is at in terms of pollution and natural-resource depletion."

Patagonia's commitment to the environment came to life in 1972 when Yvon Chouinard, the company's founder, used its new catalog to urge his fellow rock climbers to cut back on their use of pitons and strive for clean climbing. Pitons are metal spikes with eyelets at one end for securing ropes that provide support in rock climbing. Once driven into the cracks of rocks—by climber after climber—pitons cause irreversible damage to cliffs by splitting rock crevices open and leaving the surface less desirable for future climbs. A year later, the company offered office space to Friends of the Ventura River, a California group dedicated to restoring the river's watershed and its native species.

"In 1985, we started an Earth Tax," Setnicka says. "Every year, the company gives away 1 percent of our gross revenues, or 10 percent of pretax profits—whichever is greater—to various environmental groups. We want Patagonia to serve as a model for other companies. Mr. Chouinard wouldn't stay in the business if he didn't feel there was a chance to incite social change." In 1994, Patagonia's environmental grants exceeded $1.2 million and included donations to groups like the Center for Marine Conservation and Biodiversity Associates, the Wyoming organization mentioned in the section on Natural Resources, "Looking at Nature's Big Picture." With sales growing, the company expects to give away even more money in the future and receives a steady stream of grant proposals from groups in search of financial support. "We like to give grants to local grassroots organizations," Chouinard says. "We think the individual battles to protect a specific stand of forest, stretch of river, or indigenous species are the most effective at raising issues in the public's mind—particularly those of biodiversity and ecosystem protection."

Another example of a corporation acting on its concern for the environment is The Body Shop, an English firm that sells natural skin and hair care products around the world. For some years the company has conducted what it calls an environmental audit to measure its environmental efforts and impacts on an annual basis. The results are independently verified and published. "Companies must minimize their impact on the environment, and we've found that an environmental audit requires that we be honest about what we're doing both right and wrong," says Robert Triefus, vice president of communications. "Our policy is 'profits with principles.' We believe the burden of responsibility is on us, and we can't pass that responsibility on to anyone else."

The Body Shop started as a single store in Brighton, England, in 1976. In many ways, the founder of the company, Anita Roddick, was far ahead of her time when she opened that first store. "She captured the tone of what business would become in the future," Triefus says. "She predicted issues long before they came to pass—such as concern over the environment, animal testing, and social responsibility." Today, The Body Shop has more than twelve hundred stores in forty-five countries and prints its product literature in twenty-three different languages. In the United States alone, there are more than 130 Body Shop stores.

Environmental audits are conducted at every level of the business by Body Shop employees, Triefus says. The audits determine things such as how much electricity is being used, how much wastewater is produced, and how many bottles are being recycled and refilled. Other environmental issues involved in doing business are also considered in the audits. Once all the individual store audits are completed by staff members, an independent consulting firm studies the company's overall audit to ensure the facts reported are based on reality. The process is conducted in consideration of demanding new rules adopted by the European Economic Union called eco-management and audit regulations.

The company has produced guides for conducting eco-audits around the world, as well. After Body Shop stores in Hong Kong started spreading the word about their environmental audits, other Hong Kong companies took notice. A number of these companies have since bought copies of the training manual to develop eco-audits of their own, Triefus says. The Body Shop has launched its first social audit, too, which monitors issues at the company like community volunteerism, fair-trade practices, and other human-rights concerns. Company employees are encouraged to take off a half-day a month, with pay, to participate in community projects. "Real scrutiny is needed in doing these audits to substantiate the results," Triefus explains, "but by doing it on an annual basis, they can give you targets to shoot for—environmental goals to reach."

Some of the responsibility for raising corporate environmental awareness rests with customers themselves. Every person, as a consumer of the products and services offered by the business world, has an opportunity to make purchases based on a company's environmental efforts, or lack thereof. Consumers need to look beyond their personal use of certain products and consider the way the products are manufactured and packaged, as well as the overall environmental background of the companies whose products they purchase. Only by paying attention to business practices can consumers change the market by purchasing certain products and boycotting others. For instance, buying a product made of recycled materials is a better choice than buying one that simply claims to be recyclable. By making informed purchasing decisions, consumers can help nurture more responsible business practices.

FOR MORE INFORMATION

The Body Shop, P.O. Box 1409, Wake Forest, NC 27587; (919) 554-4900

Patagonia, P.O. Box 150, Ventura, CA 93002; (800) 638-6464

Tom's of Maine, P.O. Box 710, Kennebunk, ME 04043; (800) 367-8667

CHAPTER 25

MEDICINES FROM MOTHER NATURE

A number of modern medicines come from, or are synthetically based on, natural compounds found in plants growing all over the world. As more and more of the planet's ecosystems are disrupted or destroyed, plant species are dying off and becoming extinct. When some of these species vanish, they could be taking with them clues and answers to curing many diseases afflicting mankind. Fortunately, there are various attempts being made to try and locate many of the plant world's potential healing compounds before it's too late.

Shaman Pharmaceuticals, for instance, is going to its namesakes—traditional healers of the world's tropical rain forests—to discover plants it can use to develop new drugs for human ailments. The company, based in San Carlos, California, also strives to help the native peoples it contacts during these searches by promoting sustainable development of the rain forests. "We're looking for medicines used by traditional native healers, and in doing so we're working directly with indigenous peoples from all over the world," says Dr. Steven King, Shaman's vice president of ethnobotany and conservation. "Because these local people have been using these medicinal plants for years—and sometimes centuries—we can be fairly certain they are effective and don't have extreme side effects."

Using a team of ethnobotanists—scientists who study how traditional peoples use plants, particularly for medicinal purposes—the company is visiting indigenous cultures in many of the world's rain forests, such as those found in Peru and Madagascar. The company has discovered two drugs that are in the process of being approved by the Food and Drug Administration. Provir is an antiviral drug used for a broad range of respiratory infections. The drug's active ingredient was isolated in a medicinal plant that grows abundantly in South America. Human clinical trials are being conducted for the drug according to King, and Provir's sales potential is greater than $1 billion worldwide. The second drug, called Virend, is a formula for the skin to treat herpes. Shaman discovered the plant for this drug by showing pictures of herpes infections to native healers in Ecuador and the Amazon River Basin, King said. The healers simply brought out a certain plant and said that this was what they used to treat similar infections.

When Shaman approaches a group of people to start a plant search, the company first asks what it can provide for the community right away. The requests vary from a new water system to a longer airstrip to regular visits from trained dentists and physicians, King says. In the long term, the company helps the various communities it works with by providing a way to harvest products from the rain forest without destroying it. In addition, the company created the Healing Forest Conservancy in 1992 to provide more help to native people. This nonprofit conservation organization—funded in part with company profits—provides money for schools and medical services, even in areas where the company is not working. The foundation believes educating local cultures can help preserve biodiversity. "One of the keys to saving the environment is eliminating poverty," King says. "We're giving governments a reason to protect their resources, and local people a way to make a profit from nontimber products. We need to give local people and their governments alternatives to simply hacking down the forests."

Other pharmaceuticals currently in use that have tropical origins include: pilocarpine, used to treat glaucoma; quinine, used to fight malaria; and vinblastine, used to treat certain cancers. Shaman's mission is to develop a more efficient drug-discovery process by isolating compounds from tropical plants with a history of medicinal use by native peoples. The company imports all of the plant material it screens from South America, Southeast Asia, and Africa, King says, and requires that all plant collecting be done in a sustainable manner, which includes replanting in areas of intensive harvesting. Company policy also requires that each plant targeted for large-scale production must have supply sources from several countries or be able to be produced synthetically. A diverse supply minimizes the chance of extinction from overharvesting.

Shaman's drug-discovery team consists of an interdisciplinary group of ethnobotanists, pharmacologists, physicians, and other scientists. The team focuses on local peoples' recognition of common symptoms. Team members avoid terms such as "malaria" or "parasites," which might be recognized by locals and bias their responses. The goal of Shaman's method is to identify plants used by natives for certain symptoms and then collect them for analysis. Additional collections and observations are made when native healers provide health care to local people.

Of the company's seventy-six full-time employees, sixty-one are in research and development, and twenty-five have doctoral degrees. Shaman Pharmaceuticals is guided in part by a board of fifteen scientific advisors. One of the company's advisors is Dr. James Duke, an economic botanist from the U.S. Department of Agriculture (USDA). Duke has studied medicinal plants, including herbs, for more than thirty years. When he looks at the world's rain forests, he sees money, medicine, and new jobs for many tropical nations. "I have a great interest in helping tropical countries develop their medicinal plants," he says. "Two decades ago, most chemists anticipated that the majority of modern medicines could be produced synthetically. But that hasn't happened. About 25 percent of our modern medicines are still derived from, or patterned after, plant chemicals. In fact, aspirin, the most commonly used medicine in the world, originally came from a plant." In the world of medicinal research, Duke's respect for indigenous peoples' knowledge may actually save time, money, and lives by having them direct him to the right plants.

Just 2 percent of the world's rain forests have been thoroughly researched for potential medicinal plants, Duke says. Many treasures may still be hidden, while others may have already been lost forever in areas of destroyed forest. The entire world could benefit from new medicinal discoveries, and Third World countries could benefit economically from these developments. "I now have a list of one hundred existing medical compounds found in the rain forests of the world," he says. "Most of these compounds are currently produced in the First World, but most could be produced and refined in the tropics. In many cases, this would lower the cost of production and improve Third World economies, which would then take some of the local pressures off the rain forests."

First, however, an inventory of the rain forests must be completed, he said. Once these forests are fully inventoried, it will be easier for countries like Indonesia and Brazil to justify more preservation efforts. These efforts will lead to another form of economic gain, one that doesn't require the destruction of an ecosystem. Instead of causing harm, this new type of economic incentive, based on plant-harvesting for medicines, should help sustain the rain forest.

Duke relies heavily on his senses to spot plants with medicinal potential: "You can tell when a plant has a distinct odor or when its stinging nettle causes an immediate reaction. But one of the best screens is what the local people know. I like to bring back the plants they say are either medicinal or poison. Very often, the poisonous ones can be medicinal when applied appropriately."

In the world of ethnobotany, Duke is known as Uncle Sam's Medicine Man, given the nature of his work and his employer. "I brought back nine hundred pounds of specimens from China once, but most of my work has been done in Latin America," he says. "I collected about ten thousand specimens in

Panama, and about forty of those species were previously unknown to science." After a 1993 introductory trip to the rain forests on the Indonesian island of Sumatra, Duke's imagination ran wild. He saw plants that were new to him and possibly medicine. "As I familiarized myself with the plant life in Sumatra, I couldn't help but wonder how many unknown species were within my reach. These unknown plants could possibly cure cancer or other mysterious diseases." Duke hopes to start working closely with the Indonesian government to complete a thorough inventory of the country's forests.

Even one of the world's most promising anticancer drugs, taxol, has an Indonesian connection. This drug is extracted from the bark of the yew tree, which is found in the northwest United States and the mountains of India. However, another species of yew, named after Sumatra, has recently gained serious attention. "I would like to investigate this tree in Sumatra myself," Duke says. "In fact, it would be to Indonesia's advantage to have it checked out. It might be a better source of taxol than anything we've got right now. I predict taxol will soon be a billion-dollar-a-year industry." A large American pharmaceutical company, Bristol Meyers, makes taxol from a yew species found in the Himalayas, but that tree is threatened, and additional sources are necessary to sustain the development and production of the drug.

Sumatra and all of Southeast Asia are fascinating from a pharmaceutical point of view, according to Duke. "Up until now, the majority of medicinal research has focused on the Amazon region of South America. The National Institute of Health, for instance, has invested much more time in some of these rapidly vanishing forests than they have elsewhere," he says. "There's no reason to not do just as much prospecting in Indonesia and other parts of Southeast Asia."

To help Indonesia tap its medicinal potential, Duke sees companies like Shaman Pharmaceuticals playing a major role. Shaman is sincerely concerned with treating its tropical partners fairly, he says, and that helps preserve the area and improves cooperation. "I like Shaman because we both respect the people of the forests. On the other hand, my only criticism of companies like Shaman is that they're looking for silver-bullet cures—they're trying to isolate and extract only the most active plant compounds. In contrast, I believe in the synergy of plant compounds when they are allowed to work in combination, as opposed to isolating and using just one compound."

Even the wild lands and forests of America offer medicinal treasures. The Pacific yew tree is found here, and herbs like bloodroot and wild ginseng are available to the gatherer with a trained eye. To help promote the research, use,

Dr. James Duke, left, and associates inspect potentially useful and valuable medicinal plants on the Indonesian island of Sumatra. (Photograph by Gary Chandler.)

and conservation of these herbal species, Duke serves on the advisory board of the Herbal Research Foundation, an internationally recognized center for herbal research and education. Unfortunately, many herbs found in North America are being harvested at an unsustainable rate in order to serve individuals in search of alternative herbal medicines.

"The so-called silver bullets of modern medicine often kill the enemy illness at the cost of weakening the patient's immune system," Duke says. "On the other hand, herbs are gentler—they weaken the enemy while strengthening the patient." For instance, mayapple contains a promising treatment for testicular and small-cell lung cancer. In 1992, the market for this medicinal compound exceeded $275 million. "A lot of people are also using another herb called the coneflower because it's known for boosting the immune system," Duke says. "As a result, it is now endangered in Kansas and Missouri."

Additionally, the wild yam has become increasingly popular with women who think—wrongly—that when turned into a cream it is a natural source of progesterone. However, this is not the case. Another falsehood that has resulted in a threatened plant species involves the overuse of goldenseal, which is often used to alter the results of drug tests. This treatment is actually quite likely to result in a false positive result. And the herb bloodroot is effective against plaque on teeth, but as a result of its recent popularity is threatened in parts of North Carolina.

The Herbal Research Foundation promotes the true benefits of herbs, Duke explains. Its mission is to help nature meet more of the world's health-care needs. In doing so, health-care costs go down, the environment is protected, farmers are supported, and cultural diversity is preserved. Based in Boulder, Colorado, the foundation offers information on herbs that are generally recognized as safe by the Food and Drug Administration. The foundation also offers reports written by Duke regarding the availability of herbs and their prices in the United States. To date, Duke has written fifteen books on medicinal plants and their uses. He recommends the Foster and Duke *Peterson Field Guide to Medicinal Plants* for introductory reading. "Some people see medicinal plants becoming the Microsoft industry of the next decade," he says. "I hope they're right—but even more plant species will become endangered if the growth isn't properly managed."

Our future ability to use plants to cure human ailments relies on two main undertakings. The first is saving endangered ecosystems around the globe. By doing so, society will end up preserving plants that could lead to cures for an unknown number of diseases. The second is identifying plants that are helpful in combating human ailments and practicing sustainable harvesting in order to preserve the future of those beneficial species. By working toward both of these goals, we may someday actually find a cure for cancer, as well as for any number of other troublesome diseases.

FOR MORE INFORMATION

Herbal Research Foundation, 1007 Pearl Street, Suite 200, Boulder, CO 80302; (303) 449-2265

Shaman Pharmaceuticals, 213 East Grand Avenue, South San Francisco, CA 94080-4812; (415) 952-7070

Gardening can be much more than seed, soil, and plants. Here, John Jeavons of Ecology Action draws up a Biointensive Method garden plan. (Photograph by Amy Frenzel. Used with permission.)

CHAPTER 26

EFFICIENT GARDENING

Gardening is one of the most popular hobbies in the world. Through resource-fulness, gardeners can help the environment by undertaking activities such as composting and by consuming the food they grow instead of driving to the gro-cery store for fresh fruits and vegetables. Cultivating a garden also allows people to witness firsthand some of the natural systems of the earth in action—from the growth cycles of plants to the various ways insects interact with plants. Per-haps this insight alone can help people better understand humanity's relation-ship to the earth's environment.

In an effort to save water and energy, an intensive gardening method developed by the nonprofit group Ecology Action is helping gardeners around the world increase vegetable yields by two to six times over the amount pro-duced on a similar piece of land using large-scale farming practices. This low-technology agricultural system, developed by the group's founder, John Jeavons, can turn a small plot of land into a significant food-producing unit. And the sys-tem uses a fraction of the fertilizer needed for most farming.

Jeavons formed Ecology Action in 1972 as a response to the world hunger problem and to the need of many people around the world simply to feed them-selves. The California group developed and tested a small-scale food-raising sys-tem based on Chinese agricultural techniques used four thousand years ago. Jeavons started a demonstration and education garden to test applications of this intensive approach to gardening. He calls his development the Biointensive Method because of the large yields of food it can produce and has written a great deal about it. The group's effort centers on promoting the advantages of caring for the soil while growing food efficiently. The basics of the Biointensive Method are easy to learn and rely on local organic materials to enrich and improve the soil.

Because the method requires manual labor and involves low start-up costs, it is well suited to people living in poverty who could benefit from more varied diets, according to Jeavons. And it doesn't require the use of special seeds, which farmers in developing countries often can't afford. The only tools necessary are a spade and a digging fork to loosen the soil to a depth of twenty-four inches, and even these tools can be improvised. After adding large amounts of compost and organic fertilizers, the amended soil is healthier and has the capacity to grow tremendous amounts of produce. According to Ecology Action, a gardener could grow a year's supply of vegetables and soft fruits in a six-month growing season on as little as a hundred-square-foot piece of ground. After the soil is properly prepared, which does take effort, the food could be grown with an average of ten minutes of work a day. "People using the Biointensive Method not only grow more food, they also build up the soil at the same time," Jeavons says. "I think the best solution to raising enough food for the world involves more people raising their food locally on a small-scale basis."

Ecology Action has published a number of books and other materials on the Biointensive Method. The group's book *How to Grow More Vegetables* describes the basic techniques of the system. It has been read by more than 500,000 people and has been translated into five languages. The book is used by various international organizations, such as the Peace Corps and UNICEF, as a training manual for overseas nutrition work.

The Biointensive Method is now being used in more than one hundred countries. In Mexico, thousands of people are using it to grow food, thanks to the distribution of Ecology Action materials published in Spanish. In other countries, India has started a national education and training program, Kenya has trained more than thirty thousand farmers in the method, and the Philippines has developed a national Biointensive education program for all its children. In addition, Ohio University plans to start a four-year undergraduate program based on the Biointensive Method of small-scale farming.

"Sustainable agriculture involves much more than just reducing the use of chemicals," Jeavons says. "The goal of Biointensive farming is to recycle all nutrients, grow compost crops that build and maintain healthy soil, provide nutritious food for people, and integrate trees into the farm. This type of agriculture is an essential part of building sustainable communities."

Despite its worldwide impact, Ecology Action has remained a small organization. The heart of the effort is still the group's research garden nestled in the hills above the northern California town of Willits. Although the research garden serves primarily as a laboratory for soil fertility, crop production, and seed collection, the site also has been used to train teachers and practitioners in the Biointensive Method. "Our data is continually upgraded with input from thousands of gardeners around the world who are using and testing the method in their own specialized climates and conditions," Jeavons says.

Another innovative gardening idea has been successfully demonstrated by the city of Newark, New Jersey, where city residents are covering vacant lots with blooming crops. By coupling its urban gardening program with a leaf-composting project, the city has reduced landfill costs while beautifying plots of land that were once eyesores, says Frank Sudol, chief of research and program development for the city's department of engineering: "We had an ongoing problem with maintaining these vacant lots because of illegal dumping. Through the program, we've stopped the dumping, cut maintenance costs, and put the land to productive use." Nearly one thousand once vacant lots are now being used by residents—predominantly seniors—who adopt the inner-city plots, plant and maintain gardens on them, then consume the produce they grow.

The leaf-composting project is essential to the success of the community gardens because New Jersey's rocky ground makes growing gardens difficult without sufficient compost to boost the health of the soil. The city urges residents to perform backyard composting and even provides free composting units to urban gardeners. Furthermore, each fall the city asks residents who do not garden to rake their leaves to the curb. The leaves are collected and transported to Newark's composting site, where they are screened, watered, and composted in long rows. Eventually the finished product is hauled back to the urban area, where it helps the vacant-lot gardens bloom the following spring. "Rather than just hauling all that leaf material to the landfill, we use it to enrich the soil and improve the value of our neighborhoods," Sudon says. As an added benefit, more and more residents are now composting on their own and hauling the rich material to the community gardens.

An annual dinner is held each fall as part of the urban gardening program to honor the efforts of Newark's community gardeners. Before the dinner, a bus tour of the prosperous lots is conducted, and a panel of judges chooses winners in a variety of categories. Slides of the winning gardens are shown at the dinner as neighborhoods celebrate their efforts and make plans for another growing season. "We're talking about a significant amount of acreage being gardened that would otherwise sit unused," Sudol says. "The program has helped raise community spirit and involvement. More than $750,000 worth of produce is grown on our vacant lots each year."

Another gardening innovation that can be undertaken on buildings everywhere hopes to transform useless black rooftops into prosperous green gardens.

Dr. Paul Mankiewicz has designed a unique greenhouse that solves the complexities of gardening on rooftops and will provide fresh produce for residents below. By using lightweight soil, a simple steel-frame structure wrapped in plastic, and a unique planting and harvesting system, this new greenhouse could add a new facet to agriculture. "If we have a system that can purify the air and lower the amount of traffic needed to ship produce, and place that system in cities where pollution problems are serious, that is certainly a big improvement for society," says Mankiewicz, director of the Gaia Institute, a New York City environmental research group. "And by having a new center of economic activity right in the cities, it makes for an increase in wealth for those urban areas."

After several years of research and development, Mankiewicz has applied for several grants to support building a prototype greenhouse on a building adjacent to the Cathedral of St. John the Divine, New York City's fifth most popular tourist attraction. Involved in the grand proposals is New York City's recycling division, which will help supply the ton of solid waste the greenhouse will require every day. Food scraps and other waste material composted in large bins on the ground could be slurried up to the roof, where the material will be used in both the soil and the nutrient system. Composting also could take place on the rooftop, a practice Mankiewicz now leans toward. "Food-waste composting has two major advantages for a project like this—heat production and carbon dioxide enrichment," he explains. "Food waste from a hundred-table restaurant, when composted, could generate enough heat to warm a rooftop greenhouse. And the carbon dioxide generated from the composting process increases plant growth and is commonly used in greenhouses around the world."

The greenhouse's soil is the unique feature that makes the system work. Most soils are too heavy to use on rooftops, and constructing additional reinforcement to protect the building is too expensive. However, Mankiewicz has created a super-lightweight soil by using both synthetic and organic materials, including recycled Styrofoam. The Styrofoam works as filler—the purpose heavier sand and clay serve in most soils. Nourishing the crops involves a series of underground tubes linked to a controller, which delivers precise amounts of water, nutrients, microbes, carbon dioxide, and oxygen to maximize plant growth.

Covering the plants will be a lightweight steel frame covered with a thin glazing of plastic. By his calculations, Mankiewicz thinks the amount of petroleum needed to produce all the plastic used on the greenhouse would get a truckload of produce just three hundred miles down the road from California.

In the long run, this project will save much more petroleum than is consumed in its construction. For planting and harvesting, a space-saving gantry system will hover above the garden and roll over the top of the growing space. With this design, workers will be able to work from above—reaching down to the plants—thus eliminating the need for aisles and increasing crop yields by 30 to 90 percent.

On a larger scale, Mankiewicz envisions rooftop greenhouses eventually adorning the tops of shopping malls. With tens of thousands of square feet available, a shopping mall–roof greenhouse could supply all the produce for an entire community. And by increasing the profits of both store and mall owners, the technology could blossom, becoming a new green layer of the urban environment. If the prospect of widespread rooftop greenhouses seems far-fetched, it is an idea that should nevertheless be considered in areas where too much land has been gobbled up for developments such as shopping malls. The concept deserves further research and consideration.

Gardening, whether it occurs on a rooftop or an inner-city lot or a tiny plot of land in Africa, is a practice that benefits humanity. Though growing a small amount of vegetables in a backyard garden may seem a small matter in terms of the world's overall environment, the effort can help connect humans with the land. Native Americans worship the land, asking for its blessing in providing them food for the upcoming winter. This reverence has carried over into a thoughtful attitude toward all elements of the natural world, including the sun, rivers, animals, and precipitation. Modern society could do worse than to develop a similar form of reverence for its environment, and gardening is one way to engender a sense of caring for the natural world.

FOR MORE INFORMATION

City of Newark, 920 Broad Street, Room 410, Newark, NJ 07102; (201) 733-4356

Ecology Action, 5798 Ridgewood Road, Willits, CA 95490; (707) 459-0150

Gaia Institute, 1047 Amsterdam Avenue, New York, NY; (718) 885-1906

Organic gardening is a fun, environmentally friendly activity that people young and old can enjoy. Andy Lopez, in this photograph the Not-So-Invisible Gardener, has practiced organic gardening all of his life. (Photograph by Adam Rogers, courtesy of The Invisible Gardener. Used with permission.)

CHAPTER 27

PESTS AND THE
INVISIBLE GARDENER

Andy Lopez goes to work with vinegar, a bag of flour, garlic cloves, peppermint soap, and a bottle of Tabasco sauce. He may sound like a Cajun chef, but he's actually one of the world's most respected organic gardeners. He uses these items as pesticides. As time passes, more and more so-called safe chemicals are outlawed, he says, *after* society has been used as a guinea pig. That's why Lopez promotes organic alternatives through books, tapes, lectures, and radio shows.

"Peppermint soap is toxic to insects and ants," he explains. "It scrambles their senses and drives them away. I put Dr. Bronner's [a natural brand] peppermint soap in water and spray it on plants and vegetables. It's also good for many diseases, but it has to be the right concentration—about one tablespoon per gallon." Vinegar also works well in gardens. Many plants thrive on it as a nutritional supplement and quickly absorb it. When poured in a bowl, it also can attract and kill pests. "Just cover a bowl of vinegar with some lettuce leaves and place it in your garden. Vinegar attracts snails and other creatures that are predators by nature," Lopez says. "It's a good trap. If you're getting lots of the same type of pest, you probably have an infestation problem. If you're capturing a wide variety of pests, however, then your garden population probably is balanced, and that's usually a good sign. In this case, remove the bowl of vinegar. The key is balance."

Flour is another organic alternative for gardeners. When poured around the bases of plants it creates a temporary barrier against most crawling creatures—especially when it's mixed with cayenne pepper. When mixed with compost, flour forms an excellent patch for scarred trees, Lopez advises. Liquid mixtures of water, garlic, and cayenne pepper are extremely effective against various pests when sprayed directly on the plant. "It's really very simple. The philosophy is to help Mother Nature heal herself," he says.

With each passing year there is growing concern that synthetic pesticides are linked to some cancers and illnesses. A Denver-based study, for instance, published in the *American Journal of Public Health*, found that children whose yards were treated with herbicides and insecticides had four times the risk of a certain cancer—soft-tissue sarcoma. Fetuses exposed to home pest strips during the final three months of pregnancy had three times the normal rate of leukemia, and children exposed to them after birth ran twice the normal risk rate of leukemia, the study found.

Lopez has been involved in organic growing methods all his life. He grew up in Puerto Rico and Cuba, and his mother always had a compost pile and employed other resourceful gardening practices. After looking for a way to finance his education at the University of Colorado in Boulder, he started a nighttime organic gardening business. "That's how I got the nickname 'invisible gardener.' My clients would wake up in the morning and see an entirely different garden without ever seeing me."

Lopez stresses that the key to organic pest control and gardening is healthy soil. Not only are most chemicals hazardous, but they are too extreme, he cautions. To control pests organically, he warns people not to overdo it: "You don't want to sterilize the place." Gardeners need to read the labels of the products they use, he says, because many products claim to be organic or natural but aren't.

Lopez runs a business and an association of organic gardeners under the Invisible Gardener name. He stresses that the term "invisible" refers to the environmental impact of organic practices. He has written two books on the subject: *How to Heal the Earth In Your Spare Time,* and *Natural Pest Control: Alternatives to Chemical Pest Control for the Home and Garden, Farmer and Professional.* Lopez also has a radio talk show, publishes four newsletters a year, and manages an organic gardening help line and an on-line bulletin board service. He started his business in 1972 in Miami, Florida, but moved to Malibu, California, in 1984. On top of all this, the invisible gardener even makes house calls in the Los Angeles area, providing his expertise to homeowners and professionals.

Lopez's Invisible Gardeners of America is an international association that boasts more than six thousand members. The club provides information to its members on various methods of organic pest control and natural pet care. "The club provides information that promotes a healthier, cleaner lifestyle," Lopez says. "We act as a sort of 'Seal of Organic Approval.' If it's not organic, we know it, and we will tell you." Members receive four newsletters a year, a resource directory, a 10 percent discount on Invisible Gardener products such as compost, fertilizer, and videotapes, and free time on Lopez's on-line bulletin board service, which offers numerous categories, including the Organic InfoBank—a self-help resource directory. Other topics available include natural tree care, compost production, and ant control.

Lopez isn't alone in using Tabasco sauce for pest control. After innocently biting into a deviled egg covered with Tabasco sauce, then wiping away the tears, Ken Fischer tested and invented Barnacle Ban, a nontoxic undercoating for boats and ships. The undercoating is made of enamel paint with powdered chili peppers mixed in. When barnacles or zebra mussels touch it, they hop right off, because the peppers attack their nervous systems. Ships covered with barnacles have excess drag and use a lot more fuel than does a clean ship, Fischer explains.

In the Netherlands, Tabasco's power inspired a different use. When the government banned traditional pesticides, many farmers couldn't keep pigeons from eating their crops. Wil Bolhuis decided to spray his entire one-hundred-acre farm with a sixty-to-one solution of water and Tabasco sauce. The organic mixture is legal, and the pigeons left his fields alone. In addition, Dr. Ken Aldridge of Louisiana State University recently reported that Tabasco sauce can kill germs on raw seafood. His research team found that the sauce kills other harmful bacteria, such as salmonella. Other mixtures were tested, Aldridge said, but none were as effective as Tabasco.

The success of these innovations is due to the cayenne pepper in Tabasco sauce. The peppers are very high in acetic acid, which causes the burning sensation. In concentrated form, the acid can actually cause blisters. The McIlhenny Company of Avery Island, Louisiana, has produced the Tabasco brand since 1868. "We're fascinated with these new ideas," says Paul McIlhenny, vice president and secretary of the company. "Our product is a food condiment, not a pesticide or herbicide. We don't promote these new applications—but we don't oppose them either." McIlhenny speculates that red pepper may have been used in an early form of chemical warfare among ancient civilizations. He also once heard a story about a Wyoming rancher who sprayed his sheep with a cayenne pepper mixture to repel hungry coyotes.

Employing a different philosophy, a California company has developed another alternative to chemical pesticides. In an unusual "good guy versus bad guy" scenario, the company is sending beneficial insects to battle harmful ones. Nematodes are microscopic worms that can be grown and used as natural insecticides. The BioSys Company recently perfected a system of growing these organisms in a special fermentation process and packaging them in a formula that keeps the nematodes alive for five or six months. "We've been researching this idea for the last seven or eight years," says Rich Katzer, vice president of marketing. "And we are now the world leader in the use of nematodes as insecticides."

The company's products can control dozens of destructive insects that attack grass, shrubs, fruit trees, flower beds, and vegetable gardens. Grubs, cutworms, and root weevils are among the most destructive varieties. While the nematodes are lethal to these "bad" bugs, they won't harm humans, animals, groundwater, or even other beneficial insects such as earthworms, ladybugs, and honeybees.

One of the company's products, called BioSafe, even attacks fleas, providing benefits to pet owners as well as to home gardeners battling other insects. Dog and cat fleas are the most widely spread and abundant of all fleas. They have life spans of more than one hundred days, during which time they can produce thousands of eggs. To combat these fleas in the early stages of their lives, consumers simply add water to the BioSafe granules, stir, and spray the mixture on their lawns using a hose-end sprayer, sprinkler can, or tank sprayer. No special safety equipment is required.

Beneficial nematodes are part of the natural balance of the soil, where many undesirable insects spend much of their lives, Katzer says. The nematodes track their prey through various means, including following carbon dioxide and waste emissions. They enter an insect's body through natural openings and then release deadly bacteria inside the insect. The nematodes then reproduce inside the host body, creating a new generation of beneficial nematodes. When the prey insects have been controlled, the nematodes disappear through natural mortality. Unlike many chemical insecticides, a nematode application can last five or six weeks instead of just a few days, Katzer adds. The majority of the wormlike creatures are microscopic and not visible to the naked eye.

The Environmental Protection Agency has exempted these nematodes from normal registration requirements because they are harmless and naturally occurring organisms, Katzer says. Because they are not chemicals, treated land can be used immediately after spraying. One package of BioSafe contains more than 100 million nematodes and will treat about 2,500 square feet of ground. It can be found at many pet stores as well as in regional home and garden centers. BioSys also has products available for golf courses and lawn-care professionals.

"Our products eliminate or cut down substantially on the use of chemical insecticides," Katzer says. "This means far less problems with chemical residue affecting produce, farmworkers, or communities in agricultural areas. A number of large chemical companies are now interested in this type of insect control.

They realize this is the new wave of insecticides and can see that in ten years, many of their products will be outdated."

As these examples demonstrate, lawns and gardens don't need to be doused with toxic chemicals to be healthy. Using natural forms of pesticides can make a huge difference for the environment by keeping harmful chemicals off the land and out of water resources. Equally important, however, organic lawns and gardens are a healthier alternative for people, their families, and the planet.

FOR MORE INFORMATION

BioSys, 10150 Old Columbia Road, Columbia, MD 21046; (800) 821-8448

The Invisible Gardener, P.O. Box 4311, Malibu, CA 90265-4146; (310) 457-4438

WILDLIFE

Born to Be Wild

Civilization's Press on Indonesian Wildlife

Helping Africa's Wildlife and Its People

Protecting endangered species also means protecting the places they live. Here, researchers film Atlantic spotted dolphins that are the subjects of the Wild Dolphin Project. (Photograph courtesy of Wild Dolphin Project. Used with permission.)

CHAPTER 28

BORN TO BE WILD

Thanks to numerous projects striving to preserve and revive animal species around the world, more than animals are being saved. The ecosystems they live in also benefit. But are these conservation efforts enough? Today, humanity's quest for "progress," seemingly insatiable, rules over many animal species that get in its way. If that quest continues without regard for animals and the ecosystems that support them, the future of the earth could be threatened.

Although many people believe in measures aimed at protecting wild creatures are worthwhile and necessary endeavors, others think that losing a few species will make no difference in the future of the earth. However, a loss of one or more species in a given ecosystem *can* cause irreparable harm. The delicate balancing act involved in the various food chains in the natural world is well known: If one species dies off, others can be negatively affected. And if a domino effect is set in motion, whole ecosystems could potentially fail.

Although the total collapse of most ecosystems is not imminent, partial collapses already are affecting humans. For example, on the Southeast Asian island of Guam, certain species of bats have been harvested by the thousands because they are considered a culinary delicacy in Asia. Two of the island's three species of bats are now extinct, and the third species is threatened. Unfortunately, nearly half of the island's plants rely on bats for seed dispersal or pollination. The reduced numbers of bats have affected the pollination of valuable fruit plants. When the plants started dying off, the economic fallout was a devastated commercial fruit market. The future of this ecosystem remains in doubt.

With humanity expecting to inhabit the earth for centuries to come, underestimating the impact of species losses on the environment could have catastrophic results. If we don't care about one or two species disappearing, what's to prevent more from following those into extinction? And if we're not sure what will result from the elimination of various species, why take the chance that it could end up being a fatal mistake?

One mistake humanity continues to make is in the ongoing practice of animal smuggling—the removal of animals from their shrinking natural habitats for sale to the public in developed nations. Exotic birds, because they are relatively easy to take care of, remain a favorite target for smugglers and buyers. Lizards, snakes, and other reptiles also are popular, for the same reason. And various species of monkeys remain a perennial favorite, even though they require a great deal of attention and care.

"Animal smuggling is a red flag for what's wrong with humans, which is our inability to see ourselves as part of a whole system," asserts Beth Armstrong, who works at Ohio's Columbus Zoo and helps with a wildlife rehabilitation effort in Guatemala titled ARCAS. "We're removed from the overall system and need to work at connecting with the natural world. Animals are part of that natural world, but we don't see ourselves as part of it. If we fail to reconnect, I believe this will be our downfall."

Animal smuggling is symptomatic of a lack of employment options in many nations where the practice is prevalent. Smuggling can provide a living when no alternatives exist. Although a scarcity of jobs is not strictly an environmental issue, it can lead to ecological damage through destructive economic activities such as slash-and-burn farming, which destroys rain forests, and the smuggling of animals, which disrupts ecosystems. For this reason, many efforts to curb animal smuggling involve the development of environmentally benign ways for would-be smugglers to make a living.

Another way to slow or end animal smuggling involves tackling the demand side of the question. Without a market for wild birds and animals, smuggling would not take place. "For buyers, there is a certain level of ignorance—a lack of education," Armstrong says. "People see a big colorful bird and are enamored by it, but do they realize how many other birds died in order to get that one bird to market?" Many buyers of smuggled animals seemingly don't understand the ramifications of their purchases. Unfortunately, these buyers don't fit a common profile and are therefore hard to pin down and educate.

In Guatemala, a Central American country teeming with wildlife, Tulio Monterroso decided to disrupt his native country's animal-smuggling market. "I started confiscating animals from people selling them on the street. I would basically make citizen's arrests, and the government would take the animals," he says. "Then I found out the government was just sending all these animals off to zoos. I already knew I was hurting people's incomes by doing this, then I learned the animals were being sent to cages at zoos instead of cages in people's homes."

The discovery led Monterroso to found the nonprofit group ARCAS—an acronym for the Spanish of Association for the Rescue and Conservation of Wildlife. Founded in 1989, the organization's main goal initially focused on building a wildlife rescue and rehabilitation center in Guatemala's Peten region. Encompassing the northeastern section of the country near the Caribbean Sea, the Peten is biologically rich and dense with jungles and rain forests.

A steady flow of creatures now comes through the rescue center, including birds such as parrots and macaws, many types of monkeys, anteaters, and even a jaguar, which was taken from a poacher when it was just eight weeks old. Hundreds of snakes and lizards, such as boa constrictors and iguanas, also are brought to the center. To receive confiscated animals, ARCAS communicates daily with the government's National Council of Protected Areas. This agency works to subdue poachers, then turns the animals over to ARCAS's rescue center for rehabilitation. However, in ARCAS's estimation, the government has not been doing enough to stop the smuggling because of a lack of funds and concern, so the group has undertaken the task of training local police and community leaders in the Peten.

Many times animals arrive at the center diseased or injured and require lengthy stays and much care before they can be rereleased into the wild. ARCAS has veterinarians available to take care of the sick animals: A zoo in the Mexican city of Puebla has agreed to send its veterinarians to the center on a rotating three-month schedule. Not only is their expertise appreciated, Monterroso says, but the veterinarians' ability to speak Spanish enables them to deal with the local people and authorities more easily than can visiting English-speaking veterinarians.

ARCAS's first reintroduction of rehabilitated animals involved the release of thirty-six blue- and white-crowned parrots near a Mayan ruin in a protected area of the Peten rain forest. Preparation for the release began months before the parrots were returned to the wild. Two platforms, each about one hundred feet off the ground, were erected in tall trees. One platform was to serve as a release point and the other for observation. Initially, the birds were hauled up to the platform and keep in cages there and fed twice a day. After three days the cages were opened. ARCAS had no idea what to expect. The parrots had been snatched from their nests by smugglers early in their lives. None had flown or searched for food before. Nonetheless, the birds soon were flying to nearby trees and feeding on various fruits. ARCAS monitored the birds' movements for some time after the release to check on their ability to adjust to the new environment. They did fine.

However, despite ARCAS's efforts, illegal smuggling of wild animals in

Guatemala is still all too common. Combining this poaching problem with the loss of habitat due to destruction of the rain forests could spell the end to many species in the country. "We're doing all we can to rescue the animals and get them back to their natural habitat," Monterroso says.

Helping ARCAS with the effort is the Columbus Zoo in Columbus, Ohio. Through its Project Cope, the zoo has donated thousands of dollars to ARCAS along with veterinary supplies, educational materials, and hands-on technical assistance, Beth Armstrong says. For instance, when the wildlife center was being built, the zoo sent one of its maintenance staffpersons to help with construction. The zoo is also helping ARCAS promote wildlife and rain forest conservation through ongoing educational programs in Peten schools. Nearly twenty thousand coloring books, crayons, and posters of wildlife have been supplied to primary schools in the region. Several traveling "conservation suitcases," which contain puppets, books, and other learning materials, are being circulated to teachers in the Peten. More than seven hundred teachers have been trained, and curriculum about the environment and wild animals continues to expand in the school system. These efforts help the children and their parents understand the important roles animals play in the Peten's ecosystems. By understanding that the trade of wild animals could eventually harm their part of the world, Guatemalan children may grow up with a different view from past generations. Perhaps they will even work to change some of their parents' attitudes toward animal smuggling.

"Our philosophy is that zoos have the infrastructures and resources in place to provide funding for the little things that can be effective—like coloring books and crayons—and will help kids identify with wildlife and the need to save the rain forest," Armstrong says. "I believe zoos have a responsibility to get involved in projects like this and be involved in a hands-on basis, instead of just sending checks to huge conservation organizations. We feel we're a part of something that can really have a positive impact in Guatemala, and we hope to prove that zoos don't have to invest huge amounts of money to make a difference."

Although they're not perfect institutions for animal welfare in the eyes of some, zoos can make a difference in the environment by creating a bridge between modern society and the natural world. "Zoos need to take on the job of educating people about the natural world," Armstrong says. "They can be a connection to that world by serving as a resource center." Additionally, to provide balance to their operations, zoos must work in the field to help endangered animal species and boost other conservation efforts, just as the Columbus Zoo is doing.

Of course, the natural world isn't limited to species living on land. It also encompasses the sea. Over the years, the behavior of many types of dolphins has been studied. Unfortunately, the animals usually are studied in captivity—swimming around a tank at a zoo or aquarium. But an enterprise started in 1985 has been concentrating on studying dolphins in the wild to add to research already completed on captive dolphins and to learn more about protecting the environment they live in. Called the Wild Dolphin Project, the organization is looking at the life history and communication systems of a resident group of Atlantic spotted dolphins near the Bahamas.

"By letting our researchers observe them firsthand in their world and on their terms, these dolphins allow us to learn things about their social structure, communications, habitat, and many other aspects of their lives that have never before been studied in the wild," says Denise Herzing, the project's research director. "The environmental importance of our work cannot be overstated." Unless humans understand how these animals exist in the natural world, they won't be able to develop proper ideas and measures to protect them. A better understanding of the dolphins' needs and habits should help humanity avoid intrusions on the animals' lives and damage to their ecosystem.

Now in the middle of its projected twenty-year effort, the project gathers data in the field, interprets it, and communicates the results both to the scientific community and the general public. The group studies the animals in and around a shallow sandbank in the Atlantic Ocean. The exceptionally clear water in the area allows for extended periods of underwater observation and filming of the dolphins' movements and sounds in their natural world. The project's leaders believe the best way to learn about dolphin behavior is to observe the animals unobtrusively in their own habitat over many years, then to analyze the collected data. While similar work has been done with gorillas and chimpanzees, this is the only such study involving dolphins.

Although the Atlantic spotted dolphin is not technically endangered, it *is* threatened by pollution, as is all ocean life. One of the project's goals is to work for the dolphins' preservation by using the cumulative research gathered in arguments against continued ocean polluting and overfishing. If dolphins are protected, so too is their ocean environment. "As top-of-the-water feeders, dolphins and whales are vulnerable to certain types of pollution—primarily chemicals that can't be seen, an invisible threat," Herzing explains. "They store it in their fat, and mothers pass it on to their babies through their milk, which threatens their healthy development. Ocean pollution is the number-one threat to the Atlantic spotted dolphin and many other species."

Fishing, however, could be a close second. "The information and images already collected on the Atlantic spotted dolphin played an important role in influencing the tuna industry's decision to stop using dolphin-killing fishing techniques in eastern portions of the Pacific Ocean," Herzing says. "As new threats to the world's oceans and their inhabitants arise, our research will be a major factor in the struggle for their protection."

Elsewhere, a similar animal-protection project is focused on a creature that can be found floating through the air, catching updrafts, and diving on prey at two hundred miles per hour. When the peregrine falcon reached the verge of extinction due to poaching, poisoning, and habitat destruction, the Peregrine Fund came to life. The nonprofit organization helped revive the peregrine species and now works to save other raptors. When the fund was created in 1970 by Tom Cade, professor of ornithology at Cornell University, the peregrine falcon already had vanished east of the Mississippi River, and more than 80 percent of the population had disappeared in the West, says Jeff Cilek, the fund's executive director. However, thanks to the fund's innovative technology in captive breeding and reintroduction, the falcon now is making a strong recovery.

In cooperation with other organizations, the fund has released 3,700 peregrines in twenty-eight states. As a mark of the successful rejuvenation of the species, the U.S. Fish and Wildlife Service took initial steps in 1995 to remove the peregrine falcon from the Endangered Species List, on which it has appeared since 1970. Due in part to its achievements over the years, the Peregrine Fund has been asked to participate in other conservation projects in more than thirty-five countries. To date, the group has hatched and reared birds from twenty-two raptor species.

In 1992, at the request of the California Condor Recovery Team, the U.S. Fish and Wildlife Service selected the Peregrine Fund to conduct the world's third California condor breeding program. After constructing a new seventeen-thousand-square-foot facility in 1993 at its headquarters near Boise, Idaho, the fund received eight pairs of condors from the Los Angeles Zoo and the San Diego Wild Animal Park. "Young condors raised at our center may someday be released in the Grand Canyon, providing spectacular opportunities to view the largest bird in North America," Cilek says. California condors are remnants of the last ice age. They once populated a large portion of North America. Poisoning, loss of habitat, and indiscriminate shooting reduced the number of condors to an unbelievable twenty-seven by 1987, when the last free-flying condor was captured and placed in a captive breeding program. "The Peregrine Fund is a world-renowned conservation organization, and their expertise will benefit the condor breeding program," says Michael Spear, a regional director for the U.S. Fish and Wildlife Service.

Raptors such as condors and peregrine falcons are a good indicator species because of their place at the top of the food chain. As indicators, they provide scientists with a good overall barometer of the health of an ecosystem, Cilek explains. In fact, the pesticide DDT was outlawed in part because of its detrimental effect on eagle and falcon populations when the birds ingested the pesticide through their food sources. "But the main factor affecting raptors is the loss of habitat from rapidly increasing encroachment by humans," Cilek says. "Since many raptors require a large undisturbed area to survive, their conservation provides an umbrella of protection for the entire ecosystem, helping conserve other species, too."

One of the Peregrine Fund's most dramatic discoveries occurred in Madagascar, a large island off the eastern coast of Africa. In 1930, the last Madagascar serpent-eagle was presumed to have been shot and stuffed for a museum in Paris. However, in November 1993, Peregrine Fund biologists made the first serpent-eagle sighting since 1930. The biologists later trapped one of the birds for analysis, banded it for identification, and released it. The fund is using the information about the serpent-eagle to promote additional conservation activities in Madagascar. For example, due to the discovery, the largest single block of lowland rain forest left in Madagascar will be turned into a national park. Masoala National Park will protect more than eight hundred square miles of pristine wilderness and rare habitat and will become the largest protected area in Madagascar. The general plan for the Masoala National Park was approved by Madagascar's government in 1995. "The rediscovery of the Madagascar serpent-eagle played a key role in justifying the creation of the park," Cilek says.

If these animal-protection efforts continue to result in the preservation of large tracks of wilderness and other natural settings, the environment as a whole will benefit in the long run. Protecting land from human encroachment is a crucial step toward the long-term survival of numerous animal species around the world.

FOR MORE INFORMATION

ARCAS/Columbus Zoo, P.O. Box 400, Powell, OH 43065; (614) 645-3400

The Peregrine Fund, World Center for Birds of Prey, 5666 West Flying Hawk Lane, Boise, ID 83709; (208) 362-8687

The Wild Dolphin Project, P.O. Box 3839, Palos Verdes, CA 90274; (407) 575-5660 or (310) 791-5878

Indonesia, home to a diverse number of animal species, is a natural place to test progressive wildlife conservation projects. Here, an endangered Javan rhino of Ujung Kulon National Park takes a self-portrait as it participates, unwittingly, in a rhino census-by-photo poll. (Photograph by Mike Griffith, courtesy of World Wildlife Fund—Indonesia. Used with permission.)

CHAPTER 29

CIVILIZATION'S PRESS ON INDONESIAN WILDLIFE

Indonesia is one of the world's richest countries in terms of animal and plant species. It's also the fourth most populous country in the world. That combination poses a tremendous wildlife conservation challenge to this nation situated north of Australia and composed of more than fifteen thousand tropical islands. With the habitats of wildlife clashing more and more with encroaching human populations, environmental concerns are being raised over the burgeoning numbers of people and the decreasing numbers of wild animals.

One of the most endangered of all Indonesian mammals is the Javan rhinoceros. As late as the 1880s, thousands of the rhinos wandered all over the island of Java. Then the Dutch government, which had colonized the island a century earlier, decided to put a bounty on the rhinos because they were seen as pests for trampling and eating crops. A mere forty years later, the rhino faced extinction. Only a handful survived and were restricted to the wild area on the southwest tip of the island. The Dutch finally created a wildlife reserve on Java in 1921 to give the remaining animals a chance to survive. In doing so, Ujung Kulon National Park was born, although it didn't officially become a national park until 1981. The park now is one of the few undeveloped areas on the heavily populated island.

By the 1970s the rhino's numbers had finally climbed to an estimated seventy or eighty animals. Then, in 1981, a mysterious disease swept through the population, killing an unknown number of the animals. Five carcasses were found, but many more may have died. Since then, no one has had a good idea of how many rhinos still exist in the park. That is, until monitoring efforts were stepped up in the 1990s and an arduous and lengthy project set out to count the remaining animals.

To take an animal census, nets, fences, tranquilizers, and helicopters are often enlisted to trap and count the animals. However, in the Javan rhino census, a less intrusive and more efficient way of counting animals was used because trapping one of the animals is no easy task. How do you control a powerful animal that can tear through thick undergrowth like a two-thousand-pound bullet with legs? Despite their huge size, Javan rhinos move quickly and quietly through tangles of thorny vines and seemingly impenetrable walls of shrubs. By dropping their heads, the animals simulate huge wedges that push the undergrowth up and over their backs as they plow through the vegetation.

Fortunately for the international scientists and local Indonesians involved in the census effort, cameras "trapped" the animals in their native domain. The effort marked the first time photo-trapping technology had been used to conduct a census of an endangered species. Now the photo technique may be applied to find and count other endangered species around the world. "The technology had been used before to photograph scenic shots of wildlife, but in this instance we were pushing the technology to tell us how many animals were left in existence. In essence, we pioneered the operation on this scale," says Mike Griffith, leader in the joint effort between the World Wildlife Fund and the Indonesian government to count the rhinos and make recommendations on how to save them from extinction.

Much of Ujung Kulon National Park, where all the census work took place, is not virgin rain forest or jungle. Although its few mountainous sections retain their original growth, most of the flat lowlands were deforested and farmed before the area became a reserve and park. Now the lowlands—where the rhinos roam—are a mixture of bamboo thickets, palm forests, and large tracts of dense vegetation composed of jumbled mazes of thorny vines and shrubs. "The rhinos love it," Griffith explains. "If there are thorns on a plant, it usually means it's protecting something good. And the rhinos can chew right through thorns to get at the good stuff."

Griffith originally set out to place cameras systematically along a geometric grid to cover all four hundred square kilometers of the park. However, a compromise eventually was reached in which a well-used rhino trail somewhere near the originally planned grid position would be used. The park's thick vegetation would not allow for placing the cameras on a strict grid, he said, and catching animals off their trails and in the brush would be too difficult. Even so, many times machetes had to be used to cut through the thick vegetation to reach the camera locations. In all, thirty-four cameras were used in sixty locations around the park.

Griffith, a native New Zealander who holds degrees in geology and zoology from Auckland University, first came to Indonesia more than twenty years ago. After using his geology degree in the oil business for fifteen years, he made the jump to conservation and wildlife activities after producing a book of photography for Mobil Oil. He honed his skills at photo-trapping during this work on the photo book on the wildlife of Sumatra, the large Indonesian island to the northwest of Java. During his five-week journeys into the wild lands of Sumatra to photograph wildlife, Griffith slowly learned how to effectively photo-trap animals.

"Keeping the equipment operational over long periods of time is the tough part," he says. "The secret is in the little details. Termites can get into the cameras, mice chew cables, and larger animals can just trample the equipment." Over time, the secrets he learned made his photo-trapping work as foolproof as possible. And secrets they will remain. Griffith says he thinks the methods he has developed for keeping termites out of the equipment might be worth money some day.

Before starting the photo-trapping census project, Griffith needed a team to help him. The superintendent of Ujung Kulon first picked two of his park rangers to join Griffith. They, in turn, visited nearby villages until five more men were found who were willing to go on extended expeditions into the park. "This team became one of the most satisfying parts of the whole experience," Griffith says. "They showed a lot of courage, humor, and a great work ethic. In the end we were a happy, dedicated team of jungle folk." Griffith spent the first six months doing most of the work himself while he trained his team. At first the men served as porters to get the photo equipment into the jungle.

Each camera was triggered by a pressure mat that is laid along an animal trail. In addition, each camera location needed either a six- or twelve-volt battery system to operate the camera and flashes, along with a series of cables to link all the equipment together and a tripod for the camera. Fifty pounds of photo equipment was needed for each of the thirty-four camera locations. Flashes were necessary because much of the animal traffic in the jungle occurs at night. "Rhinos are like us—they hate flies," Griffith says. "So they move a lot at night when it's cooler and the flies aren't out."

After spending many days and nights out in the forest during the first six months completing initial work while training his team, Griffith gradually let the men take on more responsibilities. He became impressed with their dedication to the effort. "They became a great source of pride for me. When tours were given to visit some of the camera sites, they would dazzle the visitors with their expertise in working with the complex photo systems." After each camera was set up, the first shot on every roll of film exposed a team member standing on the photo mat with a sign showing the name of the location and holding a survey pole. The pole provided a simple way to accurately measure the height of whatever walked by because the cameras didn't move—except when some of Java's wild dogs, called *ajaks*, got hold of cameras and knocked them over out of curiosity, as happened several times during the project. Ten cameras succumbed to assorted problems and animal high jinks during the two-year period. Three were lost in the first couple of weeks alone as the team figured out how to avoid such troubles.

As animals came down the paths, side-view pictures were taken of them as they tripped the cameras. Film was collected every four or five weeks. In all, nearly six hundred rolls of film were shot during the census—a total of more than twenty thousand images of Indonesian wildlife tramping through the forest. Each shot showed a date and time of exposure so the scientists could tell if a leopard was on the heels of a pig, for example. "In the end, we got a breakdown of all the animals of Java," Griffith says. "We had a lot of pictures of wild boars, along with shots of wild buffalo, three types of deer, otters, both black and spotted leopards, and the wild dogs. Even a peacock was heavy enough to trigger a camera one time."

And, happily, Javan rhinos also were captured on film. Several times close-up shots showed the animals' faces—with huge flared nostrils—as they came near to sniff the camera equipment and triggered yet another photo. A crucial part of the study involved identifying individual animals so that a realistic count could be reached. Griffith painstakingly accomplished this through hours and hours spent with a magnifying glass hunched over a light table, poring over slide after slide. A series of different observations allowed him to mark individual animals. "After noting the date and time of day when the rhino passed the camera, the most important thing was to identify the individual," he says. "All future population calculations based on the field data depended on identifying individual rhinos."

The survey pole provided a starting point by revealing an animal's height to within a few centimeters. After that, Griffith would look at the horns of the males. They are individually unique in their shapes and sizes. He would name the horns—such as "rose thorn," "tall spire," or "tower shape"—to mark an individual. Unique scars also were a perfect way to identify a specific animal. And a rhino's unique pattern of eye wrinkles, neck folds, or even skin pores could further mark an individual animal. "Like fingerprints on a human, the arrangement of skin pores on the body of a rhino is unique, and generally the photos taken in the field had sufficient resolution to directly compare the pore patterns of one animal with those of another," Griffith explains. Of course, the animals also had to be sexed as accurately as possible. Typically, the presence of horns or external genitalia marked the males and the presence of a calf the females.

Some of the rhinos ended up being caught on film more than a dozen times in different camera locations, giving Griffith a good idea of their home ranges. One animal's territory measured ten kilometers by seven kilometers. In the end, the photo-trapping effort and subsequent calculations revealed a Javan rhino population of about forty-six animals in Ujung Kulon. There are fewer than a dozen more of the animals in Vietnam, which means the world population of fewer than sixty rhinos remains dangerously low, but at least everyone now knows roughly how many animals remain. In the first half of the nineteenth century, the Javan rhino could be found in Burma, Vietnam, Malaysia, Thailand, and India as well as on Java and Sumatra.

In a presentation to the Indonesian government, Griffith stressed the importance of security measures in Ujung Kulon National Park to ensure that no poaching takes place. Even though poaching hasn't been a problem with the one-horned rhino, its limited population must be aggressively protected in Ujung Kulon; rhino populations in other parts of the world have been hunted for their horns. Griffith also suggested starting a second population of rhinos in another part of Java when the herd in the park grows to sixty or more animals. "Absolute protection is necessary to save these rhinos," he says. "Eventually, we'll need a second population in a completely different area to eliminate the risk of a disease wiping out the only remaining herd."

The International Union for Conservation of Nature has proposed an ultimate goal of raising the Javan rhino population to two thousand animals. It's a difficult goal but one Griffith said he hopes is eventually reached over the next several decades. "I developed a real kinship with the rhinos during this work, and I'm absolutely committed to them," he says. "I hope this effort makes sure these rhinos get a chance to survive. And hopefully, it will lead to other conservation efforts, as well."

Just north of Ujung Kulon, on the large Indonesian island of Sumatra, another animal-saving effort is under way. It teaches the country's burgeoning human population how to peacefully coexist with the endangered Sumatran elephant—and vice versa. When pressed for space, these elephants are known to come out of the jungle literally fighting mad, destroying crops and homes and even killing their human neighbors. And an elephant never forgets an enemy.

"In recent years, at least forty people have been killed by elephants in the Lampung Province," says Ir. Panjaitan, director of the forestry department of the Lampung Province. "Most deaths were women and children who couldn't run fast enough to avoid the elephants."

Because the Sumatran elephant is an endangered species, the villagers face prosecution if they harm the rare animals. But in these turf battles of life and death, elephants have been killed too. The elephant training center at the Way Kambas National Park is teaching human and pachyderm how to live in harmony. Begun in 1982, the center has trained nearly two hundred elephants primarily for herd management, in which trained elephants are returned to their natural domain. These animals help control the wild herd and keep them away from civilization. This training has saved both elephant and human lives, Panjaitan says. Homes and crops have been spared, too.

Sometimes a herd of wild elephants will simply enter farmers' fields or villagers' gardens for easy grazing. The people, who are struggling for their own survival, have tried brandishing blazing torches, setting off explosions, and installing electric fences to scare the hungry animals away, but the elephants learn to tolerate all of these measures; for instance, they knock down trees to break the electric fences. For some reason, however, the elephants are reluctant to approach a water buffalo. Therefore, when possible, villagers keep a buffalo near their homes and crops. But the most effective way to manage a herd of wild elephants is to have a trained elephant as part of the herd. Usually these trained elephants will keep the herd away from villages, but if they should approach a home or village, signaled commands can be issued to the trained animal to take the herd away from a potentially deadly conflict.

The training center began with the help of two trained elephants from Thailand who were brought in to help capture the first Sumatran elephants. These Thai elephants lured the wild ones close enough for center staff to tranquilize them. Once sedated, the Sumatran elephants were loaded into a truck and brought back to Way Kambas. The smaller animals were led back in chains, secured to the elephants from Thailand. "We only capture the elephants that are between three and twenty years old," says Ir. Rusman, a veterinarian at the center. "At this age, the animals are ideal for training, and since the elephants live about sixty years, the trained ones can be utilized for forty to fifty years."

Once an elephant arrives at the Way Kambas Training Center, it receives individualized attention. "We have one trainer for every animal," says Panjaitan. "We typically have between fifty and one hundred elephants in training." Initially, an elephant undergoes a taming phase that lasts about one month. During this high-stress period one of its feet is chained to a tree, and human interaction is immediately introduced. In addition, food is withheld for a few days to teach the elephant that meals now come from its human partners. Trained elephants at the center are quickly introduced to the wild ones. The trained animals can help calm the wild ones and impart a sense of discipline to the new arrivals. For instance, when a wild elephant undergoing training does something wrong, a trained one will slap it with its trunk, Panjaitan says. Although this may seem harsh, a trained animal won't become a dead animal in this less than perfect world.

After the elephants complete their first month of training, they must learn to respond to human commands. The trainer starts by issuing verbal and visual commands to his giant student in combination with physical touch. At first the elephants don't respond to human voices, but the trainers like to talk to their students anyway. Before long, the animals will respond to just a simple signal and nod their heads to signify that they understand the command. After the daily morning workout, each animal is led into a huge pond for a swim, a scrubbing from its trainer, and a drink of water. Individualized training sessions begin in the afternoon. After a few lessons the elephants learn to kneel, walk forward or backward, and stop on command. The elephants even will learn to reach down with their trunks and pick specific items off the ground, then deliver them to trainers riding on their backs.

It takes approximately six months to train an elephant effectively at Way Kambas. In addition, it costs Indonesia's department of forestry about $3,300 to "graduate" an animal from the school. This includes capturing, vaccinating, feeding, and training. Since an adult elephant eats roughly four hundred pounds of food per day, a lot of time and money is spent just satisfying their appetites.

Elephant training centers have been established elsewhere around the world, but Way Kambas was the first to return trained animals to the wild in an effort to control herds. The Indonesian center exchanges information with other centers in Thailand, India, and Sri Lanka to help them to be more effective. Because Way Kambas has proven so successful, Indonesia has started two other elephant training centers on Sumatra.

Elephants aren't the only graduates of the Way Kambas training center. Many villages have sent representatives to attend the school free of charge and learn how to manage the wild animals. More than one hundred villagers from Sumatra and five from the neighboring country of Malaysia already have learned the skills of elephant control. Like the elephants, the human students spend about six months in training. "The villagers spend one month in the classroom and five months out on the grounds with the experienced elephant trainers," Panjaitan says.

The educational program at Way Kambas has ultimately been successful in boosting elephant populations in the Lampung province. In 1982, when the training center started operation, there were less than five hundred elephants in the area. Now, between seven hundred and eight hundred of these endangered animals roam wild in Lampung. The wild population is growing so rapidly that some of the elephants are being trained for use in agriculture and logging. Although some purists might frown on this use of an endangered species, it strengthens the bond between the struggling villagers and an animal that competes with them for food. The net result is a more harmonious coexistence and fewer deaths of both elephants and humans.

Java is the most heavily populated island in the world, and the Indonesian government has been encouraging Java's residents to relocate to other islands, such as Sumatra. This government transmigration program, which provides people with jobs, land, and living quarters, has rapidly increased the human population in places like Sumatra. Fortunately, Indonesia has curbed its population growth as it approaches 200 million people, but it will take a while to stabilize that huge population's impact on the country's wildlife and its critical habitat.

FOR MORE INFORMATION
Departemen Kehutanan (Forestry Department) R.I., Jalan Teuku Umar, Raja Basa, Bandar, Lampung, Sumatra, Indonesia

World Wildlife Fund Indonesia Programme, P.O. Box 7928 JKSKM, Jakarta, Indonesia 12079

CHAPTER 30

HELPING AFRICA'S WILDLIFE AND ITS PEOPLE

Africa is home to some of the most spectacular wildlife species on earth. The cheetah, mountain gorilla, African lion, and many other species are found only on this continent. But because of increasing human encroachment and development, it will take a major change away from current trends to save many of these species from extinction.

One person who fought for change was Dian Fossey, who devoted her life to the research and conservation of the mountain gorilla. When she accepted the challenge, she knew that greedy poachers had to be stopped from killing the extremely rare animals. Sadly, someone stopped Dian Fossey from pursuing her goals by killing her at her Rwanda research station on December 26, 1985. "When you realize the value of all life, you dwell less on what is past and concentrate more on the preservation of the future," reads part of the last entry in the famous researcher's diary. Fortunately, some farsighted individuals and organizations have followed in her footsteps and are continuing to stand up for the continent's great variety of species.

When Fossey was killed, there were just 240 mountain gorillas known to be alive in the entire world, and with Fossey's death the future of the species became even more unsure. Fortunately, her colleagues refused to be intimidated and kept up the research and fight for the gorillas' survival. As a tribute to Fossey, the American-based Digit Fund changed its name to the Dian Fossey Gorilla Fund in 1992.

"The Digit Fund originally was set up by the late Dian Fossey for the explicit purpose of protecting the mountain gorillas," says Dieter Steklis, executive director of the fund. "We carry on that work today. Because of our presence in the Virunga region [of central Africa] over the past twenty-seven years, the gorilla population has actually been increasing—largely due to our antipoaching patrols. These patrols began in the early 1970s with Fossey herself organizing and leading them. The patrols started shortly after she set foot in the Virunga

Volcano region because she realized very early on the serious threats posed to the gorilla population by people coming into the park to hunt the gorillas or other game."

Mountain gorillas are not hunted for food. Their babies are worth money in the international black markets for wildlife. Although it is extremely difficult to smuggle the baby gorillas across borders, it does happen. In 1995, four mountain gorillas were killed in Uganda as poachers pursued a nursing baby, which could potentially have brought them thousands of dollars. The poachers were never caught. This tragedy, along with three other gorilla deaths in Zaire, came at the end of a period of ten years during which none of the animals were killed. Along with the baby gorilla trade, some wildlife collectors pay native Africans for the hands and heads of mountain gorillas. This execrable practice has led to the death of more gorillas for relatively small amounts of money.

The gorilla fund employs thirty people from the central African nation of Rwanda to run antipoaching patrols and serve as trackers to find the endangered gorillas for researchers. These patrols work the high altitudes of the country's Parc National des Volcans (Volcanoes National Park), destroying snares and confronting trespassers. This area is part of the larger Virunga Volcano ecosystem in Equatorial Africa, which spills into Rwanda and Zaire. About three hundred mountain gorillas live here, and another three hundred live in a separate ecosystem farther northeast in Uganda's Bwindi National Park. (During Fossey's life, the Ugandan gorilla population had not been discovered.) This constitutes the total number of known mountain gorillas on earth. The gorillas, which can live for more than fifty years, roam the volcanic slopes at altitudes of 8,500 to 13,000 feet. Despite the civil war that began in Rwanda in 1990, the antipoaching patrols persist in their efforts while refugees, soldiers, and rebels stream through the lower portions of the park, Steklis says.

Fossey first went to Africa in 1963, where she met Dr. Louis Leakey, the legendary paleontologist. After hearing Leakey describe the need for research on great apes, Fossey eventually decided to work with the mountain gorillas. She began her research in 1966 in the former Belgian Congo but left for Rwanda shortly thereafter due to the political upheaval in the Congo. In 1967 she established Karisoke—a research station in Rwanda's Parc National des Volcans. In 1970 her studies reached a breakthrough when a young male gorilla she named Peanuts reached out and touched her hand—the first-known peaceful gorilla-to-human contact.

After intense observation over thousands of hours, Fossey gained volumes of new information on gorilla behavior, not to mention the trust of the wild groups she studied. She became particularly attached to a young male she called Digit, so named because two of his fingers were fused together. In 1977, poachers killed Digit and cut off his head and hands as collector's items. An infuriated Fossey used the incident to capture international attention. With the help of a story in *National Geographic* magazine, contributions flowed in from around the world, allowing her to start the Digit Fund and dedicate the rest of her life to the mountain gorillas.

Fossey went on to earn a doctoral degree from Cambridge University and then took a staff position at Cornell University to aid in the publication of her book, *Gorillas in the Mist,* which brought her more international acclaim. She then returned to Karisoke to continue her campaign to stop poaching and ensure the survival of the mountain gorilla. Her murder has gone unsolved, but thanks to the work of her followers, the poaching of mountain gorillas in Rwanda has virtually stopped, although threats to the population remain.

"Fossey essentially put a stop to the poaching of the mountain gorilla in Rwanda. Now the biggest problem involves poachers coming into the park for other game," Steklis says. "They set snares for antelope, but these traps still threaten the gorilla. The snares are especially dangerous to young gorillas. If we find the animals within a week of being snared, we can usually save them, but if it's longer than that, it's unlikely. When we encounter poachers, we arrest them and turn them over to the local authorities. We also have educational programs to find out why people are poaching—to stop it.

"Our intervention has worked, but our patrols have come under fire on occasion. It's a miracle that no one [other than Fossey] has been harmed or killed. The patrollers are dedicated and courageous. In fact, while being escorted by the park's armed guards, one of our patrols walked upon a group of raiders while they were dismantling the Karisoke camp. One of the raiders was shot and killed. That's how dangerous it is."

Now that direct poaching of the mountain gorilla has virtually stopped in Rwanda, conservation efforts must shift to habitat protection. This means improving the human conditions in the region. "The war has aggravated the people-related pressures and problems. Among other things, the war has plunged the local population into deep poverty, so the needs of the people around the park are far greater than before," Steklis explains. "More people are coming into the forest to hunt for meat or cut wood for firewood and construction. We basically have shifted gears to a certain extent to see what we can do to directly benefit the local population around the park. We need them to become active partners and be motivated to save the gorillas and their habitat."

The war in Rwanda also has limited what the gorilla fund can do in the region. The fund's research center, Karisoke, has been destroyed. "Because of its proximity to Zaire, the entire region is very unstable," Steklis says. "Lots of raids are being staged from Zaire into Rwanda going via Karisoke, and many of these raiding parties are fully armed—so it's become extremely dangerous to be there. This has been our greatest problem—not being able to work out of our normal base. Everything has to be done in day trips."

Gorillas are the largest of the great apes. Males can stand nearly six feet tall and weigh up to 450 pounds. Females are much smaller, weighing about two hundred pounds. Mountain gorillas spend about 40 percent of their day resting, 30 percent traveling, and 30 percent eating. A male can eat more than seventy-five pounds of bamboo a day. Scientists use photographs of the distinct nose markings to identify individuals. Fossey documented these unique nose patterns with photographs and sketches.

Three different subspecies of gorilla exist. Most zoo visitors are familiar with the western lowland gorilla, which numbers about forty thousand and is widely dispersed throughout central Africa. Eastern lowland gorillas are found exclusively in Zaire and number between three thousand and five thousand. The mountain gorilla, by contrast, is not found in any zoo in the world because zoos agreed to a worldwide ban on keeping the animals in the 1970s. Mountain gorillas did not survive in captivity, and the wild population was too small to deplete any further.

When the political crisis in Rwanda stabilizes, Karisoke will be rebuilt and researchers will return. The fund will help the country's park system rebuild its headquarters near the Virunga park boundary so it can effectively resume conservation efforts and revive the local tourism program in the park. "If managed properly, tourism is one of those resources that can help conserve the park," Steklis says. "We want to help get tourism going again because it can bring significant income to the region, which is good for the people and the mountain gorilla."

In the interim, knowledge about gorilla habitat is being attained through NASA's Space Shuttle program, which precisely mapped the entire Virunga region from space. "This information is helping us quantify how much gorilla habitat there is and how many gorillas the region can sustain. It also has helped us identify pockets in the park that are not vital to the gorilla," Steklis says. "We might be able to open up these areas for limited use by the local people—we're finding this type of effort can motivate the locals to work with us and the gorilla."

In a broader conservation effort, the African Wildlife Foundation hopes that by helping the continent's native people lead improved lives, it can create a better future for Africa's wildlife and its varied ecosystems. To this end, the foundation has developed a program with the Kenya Wildlife Service that returns 25 percent of the gate receipts earned at various wildlife parks to local communities. This money can be used for new schools, health care, agriculture, or other needs. "If you can make the parks work for the people—the people will work for the parks," says Michael Wright, president of the foundation. "By allowing local people to benefit from a wildlife park, we hope they will end up helping protect the wildlife from poaching and other problems." In essence, the wild animals become the local people's source of income, which in theory should help turn residents of local communities into staunch conservationists.

Started in 1961 by Russell Train—who also founded the World Wildlife Fund—the African Wildlife Foundation has thirty-five other field projects under way on the continent. Headquartered in Washington, D.C., the foundation maintains most of its staff in field offices in Kenya, Tanzania, Rwanda, and Uganda. "Our goal is to build the capacity and independence of African institutions and strengthen the leadership of African individuals in the struggle to save natural resources by funding conservation projects, giving aid to wildlife reserves, and promoting conservation education efforts," Wright explains. The foundation sponsors two scholarships at Kenya's Mweka College—the only En-glish-speaking institution in Africa where African nationals can be trained in natural resource management. The group also conducts numerous workshops for rangers and wardens working at various wildlife parks and reserves.

In addition, the foundation conducts a number of programs to help protect endangered species, such as elephants, rhinos, and mountain gorillas. Efforts include supporting various studies of the animals and promoting conservation measures like stopping the trade of elephant ivory in world markets. "Our work involves education and training. By protecting the wildlife, we hope to help the native people benefit from the animals," Wright says. "Our ultimate goal is to have African institutions and people take care of the wildlife. In the end, we'd like to put ourselves out of a job."

Here in the United States, several individuals started a creative effort to raise money for the African Wildlife Foundation. Rhino Chasers, an environmentally

Scott Griffiths and Rhino Chasers, *the beer that's making a difference for African wildlife.* (Photograph courtesy of William & Scott. Used with permission.)

conscious beer, is quenching thirsts across the country while helping support and promote conservation. Through an agreement between the brewer and the foundation, the William & Scott Company—owner of the brand—is sharing proceeds from Rhino Chasers sales with the organization. Having a beer named after an African animal suits the foundation and its efforts just fine. Due to poaching, Africa's black rhino population has dropped from sixty-five thousand just twenty years ago to only about thirty-five hundred today.

The idea began when Scott Griffiths, president of a Los Angeles–based advertising agency, entered his firm in the city's advertising softball league. He named his team Rhino Chasers, after a group of rugged surfers who chased down huge waves in Hawaii. Griffiths had a Rhino Chasers logo designed for the team. The logo looked like it would make a great beer label, and when a local brewer agreed to create a beer to go with the label, the concept became a reality. Griffiths then formed William & Scott, and a portion of a local microbrewery was purchased to produce the beer.

At first, the name caused some confusion. "Some customers said, 'What are you doing, urging people to attack an endangered species? It sounds like you guys are rhino bashers,'" Griffiths remembers. "Of course, that wasn't our intention at all. So we contacted the African Wildlife Foundation, and the more we learned, the more we liked it. Instead of trying to sell the world another macho beer full of sports images, we wanted to find a way to contribute something valuable to the planet, which is what enjoying the outdoors is really about anyway."

To date, more than thirty thousand dollars in revenue from the beer has been donated to the foundation. The company now ships about forty thousand cases of beer a month to its various distributors. One of the company's ads is an award-winning parody of a traditional beer advertisement. Its slogan reads, "Tastes Great, Less Killing," and features a close-up of a rhinoceros head.

"We're very grateful that Mr. Griffiths and Rhino Chasers chose to work with us," says Diana McMeekin, an African Wildlife Foundation vice president. "Whether they learn it from billboards or beer bottles, the public needs to know what's happening to the wildlife of Africa. The only way change can come about is through increased public awareness."

Although the organizations discussed in this chapter can help set up a foundation for success, the future of Africa's wildlife and environment lies with the continent's people. The Dian Fossey Gorilla Fund and African Wildlife Foundation both are working with native people to teach them about different ways to look at their environment and wildlife. These are certainly commendable endeavors; however, wars, poverty, and an exploding population base must be dealt with by many African nations before the people and wildlife of Africa achieve true stability.

FOR MORE INFORMATION

The African Wildlife Foundation, 1717 Massachusetts Avenue, Suite 602, Washington, D.C. 20036; (202) 265-8393

The Dian Fossey Gorilla Fund, 800 Cherokee Avenue, Atlanta, GA 30315; (800) 851-0203

William & Scott Company, 213 Main Street, #250, Huntington Beach, CA 92648; (714) 374-3222

OUR COMMON FUTURE:
WHAT YOU CAN DO—A SUMMARY

We hope this book has conveyed a sense of urgency concerning the need to protect our environment as well as a sense of hope for its future. Plenty of damage has been done to the earth, and more is inflicted every day. The cumulative threat to life as we know it is real, and our children and grandchildren stand to inherit the consequences.

Will they benefit from the medicines hidden inside undiscovered plants? Or will those medicinal plants be lost with the destruction of the planet's rain forests and old-growth forests? Will our children's health suffer from growing amounts of pollution in the air, water, and food supply? Will they be able to see wolves or grizzly bears in the wild, or know that a multitude of other species—now at risk of extinction—still roam the earth? We of course hope the answer to all of these questions and many more like them is yes.

But it will take some work.

Fortunately, as the dozens of people, projects, and organizations featured in this book reveal, there is reason to be hopeful. Their efforts, and thousands more like them, are working to ensure that a safe and healthy environment prevails in the future. If people like gorilla researcher Dian Fossey and Brazilian environmentalist Chico Mendes can sacrifice their lives for their environmental beliefs, the least the rest of us can do is make informed decisions about the products we buy, the natural resources we use, and the politicians we elect.

Going one step further, we can join in any number of the efforts discussed in this book. Nearly one hundred phone numbers and addresses are just a few pages away. Each one could get you involved in an effort that is making a difference in the future of our environment. By taking part in any of them, you will add your voice to the overall movement toward environmental excellence on a global level.

Another way to become involved in helping the environment entails leaping into cyberspace. There is a wealth of information about environmental issues on the Worldwide Web. The Web is part of the Internet—what has become known as the "network of networks" in the computer world, or the information highway.

If indeed knowledge is power, as Francis Bacon once stated, then the Internet and its Worldwide Web should provide everyone from fledgling environmentalists to hard-core ecologists with plenty of power with which to venture forth in numerous efforts to protect the earth. And apart from cyberspace, don't forget the wealth of knowledge found by the printed word. Many of the organizations mentioned in this book publish magazines or newsletters with plenty of helpful information. As well, there are numerous environmental periodicals awaiting interested and motivated library visitors, and new books on the environment are continually being published.

Yes, the environmental movement is about saving whales, trees, and open pieces of land, but its greater message is often missed in those details, for it's really about preserving a world that will sustain life in all its forms. All of the people in this book are working toward that singular goal, even though they go about it in many different ways. Some focus on certain species of animals, others concentrate on entire ecosystems. Some people work specifically on combating air pollution while others center their efforts on stopping the use of harmful chemicals. But in the end, they all are working to protect the earth and its natural environment. Fortunately, they are joined by millions more people all over the earth. Are you among them?

INDEX